Cloning and the Constitution

Cloning and the Constitution

An Inquiry into
Governmental Policymaking
and Genetic Experimentation

Ira H. Carmen

The University of Wisconsin Press

Published 1985

The University of Wisconsin Press
114 North Murray Street
Madison, Wisconsin 53715

The University of Wisconsin Press, Ltd.
1 Gower Street
London WC1E 6HA, England

First printing

Printed in the United States of America

For LC CIP information see the colophon

ISBN 0-299-10340-4

To my children,
Gail Deborah and Amy Rebecca

Contents

Preface

This book is an essay on the nature and scope of the American Constitution. It emphasizes especially the fundamental law in modern context, notions of "right conduct" providing coherence to and receiving nourishment from the cultural values of our time. The subject of analysis —the cultural activity addressed—is genetic engineering, not social engineering; but the locus of analysis is political science, not biological science. If this research bears a distinctive look, it is only because contemporary scholarship has accorded sparing attention to constitutional norms as tools of social analysis and consensus-building. I shall argue that the key to understanding recombinant DNA experimentation as appropriate scientific mission lies in constitutional theory and constitutional practice.

Cloning and the Constitution is actually the culmination of a process which began more than a decade ago when I first undertook to explore the parameters of a constitution, and more specifically, the United States Constitution. At that time, the visible questions of concern involved the Warren Court, protest movements in the streets, and Watergate. These ruminations led me to write *Power and Balance,* a constitutional politics overview and interpretation of American national government, its structure and function. In the middle 1970s, the controversy over recombinant DNA investigations began to attract national headlines and, quite frankly, caused my consciousness to bolt upright. The Constitution seemed to run like a thread throughout this dialogue, but practically nobody was talking about the connection. Certainly political scientists weren't talking about it. The more I reflected on the cultural dimensions of current scientific innovation, the more I came to believe that political science inquiry had overestimated rather considerably the importance of judicial decisionmaking in the process of constitutional declaration and implementation. Equally important, it now seemed to me, were the broader questions of what role the Constitution was playing and what role it should play in mediating the great issues of social policy which today arrest the nation's attention.

That the dynamics of constitutional interplay as I conceptualize that notion can undergo dramatic substantive shifts is well illustrated by noting the most recent developments in the gene-splicing controversy. With the exception of scattered citations and updates, my cut-off point for in-depth research on this project turned out to be May 1983. Yet, as I write these paragraphs in January of 1985, the most visible issue on the recombinant DNA docket was of little notoriety fifteen months ago: to what degree can or should the federal government regulate the deliberate release of contrived mutants into the environment? On May 16, 1984, Judge John Sirica shocked the scientific community by enjoining University of California–Berkeley researchers from spraying potato plants grown on open acreage with an artifact bacterium designed to mitigate frost damage. Sirica held that a full-blown hearing must be conducted on the question whether university biologists were compelled under federal statutory mandate to obtain an environmental impact estimate before they could proceed in these unique circumstances, even though duly-constituted authorities at the National Institutes of Health had followed the usual clearance procedures in assessing safety standards. However, his order was on its face inapplicable to bioengineers employed in the nation's industrial firms, because their investigations are not supported at taxpayer expense. Judge Sirica's decision—which, if upheld at trial, would seemingly require all recombinant DNA faculty researchers in the United States to stay their hands outside the laboratory enclosure until NIH personnel developed appropriate environmental impact reports—implicates extraordinary constitutional ramifications, many of them unarticulated by current media coverage. Among these are the following: (1) Assuming, as I will argue, that genetic engineering in the form of fundamental research merits First Amendment free expression coverage, to what extent can courts issue injunctions (either with or without the assistance of current statutory language) abating such experimentation in an open field? (2) What difference if the studies are being conducted in the curtilage garden of the scientist's residence laboratory? (3) What difference with any of the above where the project is publicly sponsored? (4) Do government law enforcement personnel have carte blanche to confiscate without warrant "free expression" clones being tested in open fields? (5) What difference if the "search and seizure" takes place in a curtilage and/or is taxpayer-funded? (6) How do the norms and aspirations of American political life help us evaluate legal structures which permit censorship of gene splicing as quest for truth yet provide no comparable constraints for profit-seeking biotechnology? (7) What difference for generally accepted notions of public consent and research autonomy if these prior restraints are orchestrated by the EPA rather than by the NIH? I shall not attempt answers to any of these inquiries here.[1] But I believe that the following account provides the frames of reference necessary to resolve all of them consistent with both scientific freedom and legitimate citizen concern.

My acknowledgments are of three kinds. First, I should identify my intellectual debts. In the social science community, the thinkers who have had the greatest impact on my teaching and writing are Edward S. Corwin, Karl N. Llewellyn, Robert G. McCloskey, and C. Herman Pritchett. It was never my pleasure to study under any of them, and yet, as surely as they authored the books and articles sitting on my shelves, I did study under them. There is a wisdom linking up their principal contributions which is discernible only after considerable reflection and inquiry. Among those who have written on molecular biology, I must mention James D. Watson, for *The Double Helix,* with all its chattiness, conveys powerful messages, many of which the practitioners of contemporary political scholarship too often overlook. Scientific enterprise, we need to relearn, is at least as much artistic impulse as it is the rigorous application of set procedure.

Second, I want to mention my considerable personal debts. My wife was an extraordinary source of assistance throughout the development and execution of this project, as she always has been, while my parents also provided me with much-needed support. I wish to thank as well the many members of the New Haven community who went out of their way to help me while I was in residence at Yale Law School carrying out the survey research aspects of the investigation.

Third, I owe obligations I can never repay to those who have made publication of this study a reality. These include Gordon Lester-Massman, the Acquisitions Editor of the University of Wisconsin Press, C. Herman Pritchett of the University of California at Santa Barbara, and Waclaw Szybalski of the McArdle Laboratory for Cancer Research at the University of Wisconsin. Professors Pritchett and Szybalski strongly supported and reaffirmed my faith in the research, while Gordon Lester-Massman bent every energy to convert manuscript to book. I

1. Commentary on late-breaking developments relevant to this study as of May 1985 is presented in Ira H. Carmen, "Bioconstitutional Politics: Toward an Interdisciplinary Paradigm," paper presented at the American Political Science Association annual meeting, New Orleans, August 1985.

must mention as well my editor, Carolyn I. Moser, whose efforts improved considerably the mode of presentation, and my typists, Susan Monte and Eileen Yoder, whose labors made it possible for me to surmount ever-threatening deadlines.

Introduction

Cloning! It is a new word in our everyday vocabulary, a word that carries a touch of the romantic, a touch of the forbidden. As a matter of fact, it is a very old word which, derived from the Greek, refers to propagating plant life from a cut twig or "clone." And in the technical jargon of contemporary biological science, the term can denote any one of several conceptually different processes: asexual reproduction; the creation of identical entities; bringing to life new individuals from small segments of parent forms; and splicing sections of DNA into a cloning vector. Early in the 1970s, researchers contrived means to accomplish the last of these, to recombine genetic materials from different organisms through molecular processes, that is, to clone new life forms in the sense that adding and subtracting certain characteristics from many thousands of preexisting traits would yield unique mutant specimens. At once, thoughtful, knowledgeable people could sense the potential for good and the potential for danger inherent in such enterprise. We would unlock the secrets of genetic expression and organization; we would mass-produce resources precious to the medical community. But perhaps we might also spread new diseases, fracture the delicate ecological balance, do that which some say only nature or God should do.

And so has unfolded the "recombinant DNA debate." Philosophers and theologians argue the merits in ethical terms; scientists discourse over safety implications; venture capitalists ponder the profitability of "industrial gene-splicing"; academicians question the extent to which their universities should be seduced by the lure of "DNA dollars"; legal scholars compare briefs on whether cloning is a constitutionally protected liberty; and policymakers frame innovative regulatory mechanisms to bridle this new thrust in human ingenuity. Regarding most of these aspects, the media and the world of scholarship have provided adequate coverage, though this study must devote at least some attention to all of them.

But what of the gene splicers themselves? What are their thoughts on the recombinant DNA debate? How has it affected their attitudes and work ways? How do they react to the political influences brought to bear on them, to the legal guidelines that so often constrain their inquiries? And what of the duly constituted authorities most responsible for overseeing this research—the biosafety committees in place at every institution of higher learning where cloning is carried on? How do they perceive and structure the assorted responsibilities delegated to them? How do these governmental stewards, many of them leading scientists as well, interact in the decisionmaking process with their gene-splicing colleagues? What views do they maintain on this great debate?

This work addresses these questions, charting the legal-political universe in which recombinant DNA investigations are conducted at many of this nation's leading academic research centers. More specifically, the book, employing as a primary methodology face-to-face interviews with selected samples of both cloners and their research gatekeepers, endeavors to describe, analyze, and evaluate the professional attitudes and behavior of these actors from the standpoint of social science theory.

The recombinant DNA debate, quite obviously, is not unique among public policy issues. And certainly it has arisen within a context of shared beliefs. Some of these understandings concern the nature and role of science in our culture, while others center on the relationship between scientific freedom and traditional standards of appropriate behavior. As the purpose of this study is not to present a definitive account of the gene-splicing dialogue, it eschews attempts to characterize many of these expectations and the social forces which have prompted them. The analysis will emphasize, rather, the manner in which science, politics, and law have intersected to influence that dialogue; and these intersections will be especially revealing should they bespeak long-standing, highly prized convictions toward basic questions. Moreover, the most important among them may well have achieved acceptance as *constitutional values,* by which is meant not only explicit standards of right and duty enunciated in our fundamental law and in judicial opinion, but also the functional rules, actions, and expectations undergirding our body politic.

The manner in which the term "constitution" is here employed requires further elaboration. There are no cases in the Supreme Court literature which assign to cloning the status of protected behavior. Nor are there cases which refute the proposition. But as Chapter 1 is designed to show, there is a rich fabric of constitutional understanding which informs the relationship between government action and scientific enterprise. One can hardly develop constitutional consensus respecting public controls on genetic engineering without appreciating fully the shape of these historical patterns. Moreover, as Chapter 2 is designed to show, there are judicial doctrines and determinations providing ample support for the conclusion that cloning is a form of expression, subject to constraint, therefore, but only within prescribed bounds. And, as Chapters 3 and 4 are designed to show, the entire gamut of government/cloning oversight has raised a myriad of constitutional perplexities, with assorted interests pro and con brandishing their own sets of procedural and substantive givens. At this stage of analysis, the Constitution truly becomes the "Living Constitution," for basic legal principles are here seen as expressions of public opinion and attitude, the document becoming in our hands a political instrument for public debate and public policy accommodation. I submit that it is not simply the policies themselves which merit constitutional scrutiny; it is as well the clash of interests, the clash of competing expectations toward the use of political power. Taken together, these are the stuff of constitutional politics, and the norms respecting constitutional propriety which structure the debate as well as the norms which receive vindication in the marketplace as standards of unquestioned rightness compose significant aspects of the Living Constitution.[1]

To summarize: The book's essential themes are political science inquiry, government policymaking, and genetic experimentation. But more than that, it endeavors to dramatize the inevitable tensions between science as personal liberty and law as necessary social adhesive. It is, in short, a portrayal of reciprocal constitutional relationships.

1. The term "Living Constitution" was first employed as a full-blown analytical concept (rather than as a figure of speech) in Ira H. Carmen, *Power and Balance* (New York: Harcourt Brace Jovanovich, 1978). At that time, I enclosed the expression in quotation marks, but I am now satisfied that it has received sufficient usage to stand unaided.

Cloning and the Constitution

1 Science in American Constitutional History

The building up of scientific enterprise in the United States is a story often told. The nexus between constitutional values and this development, on the other hand, is eminently worthy of our attention. A perusal of the standard texts in American constitutional history, to be sure, often makes one wonder whether anything scientific has ever had impact upon our fundamental law and vice versa.[1] But relationships often stand or fall on how terms are defined. In this case, everything depends on what is meant by "constitutional history" and "science."

It has been noted that constitutional historians address "the origin and development of all the principal institutions, practices, customs, traditions, and fundamental legal ideas that go to make up the whole body" of today's Constitution.[2] I would say this and more. A missing ingredient is the notion of attitude or expectation. The American Constitution consists not only of governing norms but also of perceptions, understandings, and degrees of affinity manifested by significant interests toward those norms. That is, the opinions of leading decisionmakers, key groups, and citizens generally regarding the value to be accorded provisions in our governmental rule structure may themselves be aspects of the Constitution and, hence, staples of American constitutional history. For the Living Constitution is law, is custom, is usage; moreover, it is an institution, a corpus of behavior patterns energized by deep-seated beliefs regarding the role of statecraft and the bounds of constitutionally sheltered freedoms. It is not enough, then, to know the law or to know

1. For example, Kelly and Harbison's outstanding text contains practically no discussion of science. In fact, the word fails to appear in their very complete index. See Alfred H. Kelly and Winfred A. Harbison, *The American Constitution,* 5th ed. (New York: Norton, 1976).
2. Ibid., 2.

3

4

the rule; one must also know why people live the law and how they live it.[3] And so the manner in which salient persons and publics have conceived relationships between science and the fundamental law could well be the essence of constitutional history, no matter how these interests may have defined science.

The Document Speaks: Patenting for Science

The word "science" appears but once in the United States Constitution. Among Congress' delegated powers listed in Article I, Section 8, is the authority "to promote the Progress of Science and useful Arts, by securing for limited Times to Authors and Inventors the exclusive Right to their respective Writings and Discoveries." And, of course, the Constitution also vests in Congress discretion to enact legislation "necessary and proper for carrying into Execution" all enumerated grants including this one.

What does the language mean? Upon first impression, the clause divides rather neatly into three pairs of concepts: science-useful arts, authors-inventors, and writings-discoveries. This triad seems to invite sequential integration, but impressive stylist though Gouverneur Morris surely was,[4] the terms "science," "authors," and "writings" do not fit well in this context. True, authors write rather than discover, and inventors discover rather than write. Still, science goes much better with inventions and discoveries, while writers tend to be considered artists rather than scientists. And what are we to make of the science–useful arts distinction? Why does it not read "useful science–useful arts" or "science–arts"? What public policy preferences would lead the Constitutional Convention to give scientists broader exclusive rights than artist-authors?

As Justice Holmes might have put it, an ounce of constitutional history is worth a pound of constitutional philology. For purposes of understanding what is patentable subject matter, the key word is not "science"; it is "inventor." We learn that as early as the sixteenth century most patents issued under English law were tendered to corporations for the purpose of shielding "manufacturing privileges."[5] Later, industrial patents designed to protect the value of inventions took hold. Across the Atlantic, this practice became commonplace around the time of the Revolution. It turns out that the patent clause in the U.S. Constitution was merely another feature of federalist strategy. That is, the

3. See generally Ira H. Carmen, *Power and Balance* (New York: Harcourt Brace Jovanovich, 1978).

4. It is well established that Morris wrote up the Constitution in final form.

5. E. Burke Inlow, *The Patent Grant* (Baltimore: Johns Hopkins Press, 1950), 13.

dominant economic interests which played such a prominent role at the Philadelphia assembly wanted to promote "useful inventions" in aid of business enterprise just as they wanted to insure the obligation of contracts for similar purposes. Henceforward, states could no more issue patents than they could impair the sanctity of debtor-creditor agreements.[6] The root idea, then, was to construct a viable, unfettered national economic order grounded in bold, assertive private property interests.

What is a patentable invention or discovery? Congress defined these constitutional terminologies in 1793 to include "any new and useful art, machine, manufacture or composition of matter." A patent, when received, was to run for fourteen years, during which period inventors retained an "exclusive property" in their artifacts.[7] Since that time there have been three statutory changes of note. In 1870, the lifespan for patents was extended to seventeen years. In 1952, the word "art" was removed, and in its stead was inserted the word "process." Finally, also in 1952, the criterion of nonobviousness was added to the "new and useful" catalogue of relevant adjectives. No doubt the inclusion of processes as applicable subject matter plugged a considerable loophole, but the deletion of "art" bears witness to the definitional problems recited earlier. Whatever "useful art" may have meant to the Founders, in our time the phrase seems somewhat at odds with the theme of patentable inventions. As for nonobviousness, the Congress was here trying to distinguish between improvements wrought by genuine talent and those accomplished through ordinary skill in the craft.[8] After all, it was one thing to obtain a patent on the telegraph, the airplane, the transistor, and the laser, but it might be quite another thing to receive seventeen-year exclusive rights to a vacuum-packed container, a railroad brake, a felting machine, and a porcelain door knob.[9] To sum up Congress' intent, patents shall be extended to "anything under the sun that is made by man,"[10] provided it be original enough and practical enough.

Where does all this leave "science" as set forth in the Constitution? The Supreme Court has never defined the word; indeed, the justices have construed other passages in the patent clause only on rare occasions.[11]

6. Ibid., 46–47. In *The Federalist*, no. 43 (New York: Modern Library), 279, Madison also talks of the patent provision as a boon to "useful inventions."

7. 1 *Stat.* 318.

8. Inlow, *Patent Grant*, chap. 6.

9. Ibid., 137–38.

10. S. Rep. No. 1979, 82nd Cong., 2nd sess., 5 (1952); H.R. Rep. No. 1923, 82nd Cong., 2nd sess., 6 (1952).

11. Instance: A trademark, it has been held, is neither a "writing" nor a "discovery." Trade-Mark Cases, 100 U.S. 82 (1879).

6

That is why Congress' sentiments here are so significant. But surely "science" must mean—and must have meant in 1787—something more than invention. Had not Madison heard of Archimedes? I shall hypothesize, then—and attempt to demonstrate in the next section the truth of such hypothesis—that the phrase "to promote the Progress of Science and useful Arts" is in the nature of a preamble, setting the stage for the particulars that immediately follow. As the Constitution itself was conceived and molded to achieve certain broad but specifically enumerated objectives, so Congress' power to patent inventions and copyright forms of words would nurture larger scientific and artistic interests, whatever those might be.

In point of fact, a few intellectually spirited members of the Court have volunteered commentary regarding the parameters of scientific investigation within the context of the *statutory* framework. Justice Douglas, writing in 1941, claimed that an artifact was unpatentable unless it bespoke "the flash of creative genius."[12] In his belief, bona fide innovations come only from quantum leaps of the fertile imagination. To this, Justice Frankfurter rejoindered: "The discoveries of science are the discoveries of the laws of nature, and like nature do not go by leaps."[13] The debate is somewhat oversimplified not only because the jurists failed to demonstrate understanding, based on empirical evidence, of what scientists actually do, but also because they lumped together uncritically the scientist's work ways with the inventor's work ways. But the discussion does anticipate a *political* question of considerable relevance: To what extent must government regulation of science be tailored to an appreciation of scientific labors?

In 1980, the patent grant became a leading issue in the recombinant DNA debate. Ananda Chakrabarty, a biologist employed at General Electric, had cloned a new form of *Pseudomonas* bacteria. These efforts stemmed from his earlier discovery that plasmids—i.e., nonchromosomal genetic material—conferred oil-degradative capacities upon certain microorganisms. While the genus *Pseudomonas* possessed no such characteristics, it proved to be a serviceable host vector through which four alien plasmids could be recombined, each section being charged with the responsibility for fragmenting a discrete oil component. Financial prospects for General Electric seemed bright indeed. Armed with exclusive access to Chakrabarty's bacterium, the company could now efficiently dispatch crude oil spills polluting the planet's lakes and oceans. But were the new life forms patentable?

By a vote of 5 to 4, the Supreme Court held that Chakrabarty's micro-

12. Cuno Corp. v. Automatic Devices Corp., 314 U.S. 84, 91 (1941).
13. Marconi Wireless Co. v. U.S., 320 U.S. 1, 62 (1942).

organism constituted a "manufacture" or a "composition of matter" within the meaning of the congressional definition and that, therefore, the patent should issue.[14] Commencing from the premise that capacious legislative objectives deserved generous construction, the Court admitted that the statute was never designed to cover "laws of nature, physical phenomena, and abstract ideas." Specifically, Congress did not intend either for Einstein to patent his theory of relativity or for Freud to patent his id-ego-superego trichotomy. Nor did the law's scope include newly found natural botanical, zoological, and geological specimens. These discoveries, the Court reminded, were "free to all men and reserved exclusively to none."[15] The issue, then, was merely whether the *Pseudomonas* were natural or contrived, not whether they were animate or inanimate. As they were man-made, they were patentable.

The dissent's approach was markedly different. Of course science should be encouraged, but patents were essentially monopolies. Courts must not extend the applicable language to uncharted areas absent overt legislative approval, especially given demonstrable public concern.

The case presents several noteworthy features for our purposes. For one thing, the two blocs fought hard to snatch the cloak of judicial modesty from one another. The minority argued that it was up to Congress to provide clear guidance, and the greater the degree of social consequence the greater the need for legislative oversight. But for the majority, even this approach constituted undue assertiveness. The Court, it found, was without competence even to assess whether the recombinant DNA debate was much ado about something or nothing. Let Congress hold hearings on cloning, the opinion ran; the judiciary's task was the "neutral" chore of reading the law as written.

Quite likely, however, the makeup of the two groupings conveys more information than this fencing over interbranch role-playing. The majority comprised five Republicans and no Democrats, while the minority consisted of three Democrats and one Republican.[16] Looking to the historic mainsprings of partisan ideology, we find that the patent clause received its staunchest support at the turn into the nineteenth century from the nationalist-business class, the faction that would shortly underlie the Federalist party. To these mercantilists, a patent was akin to a protective tariff; both would foster economic innovation and competence at home. All doubts should be resolved in favor of investment, of property claim, of patent rights. That is a credo which modern-day Republicans

14. Diamond v. Chakrabarty, 447 U.S. 303 (1980).

15. Ibid., 309.

16. For granting the patent: Burger, Blackmun, Stevens, Stewart, and Rehnquist. For withholding the patent: Brennan, Marshall, White, and Republican Powell.

can endorse without embarrassment. To the "liberal" Justice Douglas writing forty years before *Chakrabarty,* however, the primary consideration was not what would spur invention but, rather, what would promote the public welfare. This was the philosophy of *Munn v. Illinois*[17] and Jacksonian democracy. Property interests must yield, he thought, where legislators find in reason that these interests trigger antisocial consequences. Patents should issue—and, hence, vest in the discoverer immunity from competition—on those few occasions when the public itself would be the principal benefactor. For that to occur, he concluded, invention must spur dramatic improvements in scientific understanding and technological know-how.[18] Clearly, Douglas' approach is consonant with contemporary Democratic party orientation, and provides the jurisprudential foundation for the minority position.

What the *Chakrabarty* case tells us is that cloning particular life forms can qualify as "invention" under the patent act. Recombinant DNA researchers may or may not investigate the laws of nature, but they all construct artifacts. Who can blame them should they procure monopolies when the law permits. It is not only good business; it is a badge of creativity, demonstrating both genuine originality and *utility*. Does this reward system bear witness to "science" as a force for enlightenment, worthy even of constitutional specification? To that issue, we now turn.

The Intent of the Framers: A Tale of Two Scientists

Constitutional historians place great value on unearthing what the Founding Fathers meant by a particular word or sentence. But an alert citizenry undoubtedly understands why the search can be of vital importance. What is "due process"? What is a "cruel and unusual punishment"? If we knew the Framers' intentions, it would at least provide a context for today's clash of definitions.

Oftentimes, though, the world of scholarship cannot help us. There were no speeches delivered before the Philadelphia gathering, diagramming the possible dimensions of scientific inquiry. Moreover, we must ask who are meant by the "Framers." Commentators generally consider Hamilton a Founding Father, yet he rarely attended the constitutional meetings. And should not delegates to the ratifying state conventions qualify as "framers"? For purposes of this analysis, I shall define the term to include any well-known figure of the day whose expertise in a given area was such that in all likelihood the typical Philadelphia participant would have searched him out for guidance in defining and interpreting proposed constitutional language. Without any doubt, the leading

17. 94 U.S. 113 (1877).
18. Inlow, *Patent Grant,* 148–50.

American authorities on science at that time were Thomas Jefferson and Benjamin Franklin. Franklin, in fact, did sit as a delegate, while Jefferson, because he led the fight for a Bill of Rights, must be considered on that score alone a Founding Father. Certainly if anyone knew what "science" meant in this country in 1800, it was these two luminaries. Their words and deeds provide the necessary instruction.

Jefferson and Franklin are a study in contrasts. Politically, they were much alike, leaning strongly toward democratic values. But with respect to thinking and doing science, comparisons become difficult. The fact is that while Jefferson spent a great deal of time and energy extolling the virtues of science, he actually did precious little of it. Franklin, on the other hand, wrote sparingly about science as a discrete activity, but his contributions to the discipline were extraordinary.

Even those who have spent a lifetime investigating Jefferson approach their subject warily. The man who, as president, at once called himself a "strict constructionist" yet purchased on his own authority the Louisiana Territory, never formulated a complete, internally consistent political theory.[19] The same may be said about Jefferson the scientist.

First, we have his writings and, if one will, musings. To say that he placed science on a pedestal would be an understatement. Freedom of the mind meant for him, above all else, freedom to work in science, because from science we gain happiness and yet more freedom through understanding.[20] Science was reason, the quest for truth, the preserve of civilization's explorers.[21] The scientific mission must be construed expansively,[22] encompassing within its sweep poetry, criticism, rhetoric, history, mathematics, and government as well as chemistry, physics, and biology.[23] But Jefferson could focus on more traditional disciplinary pursuits when he so chose. For him, it was Joseph Priestley, the discoverer of oxygen, who stood at the "pinnacle" of science.[24] And he had grave doubts about medicine's standing as scientific enterprise, because physicians relied unduly upon hypothesis to the detriment of "sober facts," "clinical observation," and "an intimate knowledge of the human body."[25]

19. Richard Hofstadter, *The American Political Tradition and the Men Who Made It* (New York: Knopf, 1949), 23.

20. Karl Lehmann, *Thomas Jefferson, American Humanist* (New York: Macmillan, 1947), 129–30.

21. Ibid., 206.

22. Thomas Jefferson to Elbridge Gerry, reprinted in Saul K. Padover, ed., *A Jefferson Profile* (New York: Day, 1956), 112.

23. Jefferson to Peter Carr, ibid., 239–40.

24. Jefferson to Joseph Priestley, ibid., 127.

25. Jefferson to Casper Wistar, ibid., 162–64.

Still, Jefferson's own work was in the "mechanical arts," not at the level of "high abstractions." He loved to compile data, to calculate, to tinker; and his results were astounding. He invented a hempbeater, a leather buggy top, a swivel chair, a dumbwaiter, and a plow.[26]

Jefferson, it has often been pointed out, authored the nation's first patent law, and this fact allegedly provides evidence that current legislation was designed primarily to foster "new knowledge."[27]In truth, Jefferson had very mixed feelings about patenting anything. Somewhat surprisingly, the Hamiltonian rationale did not strike him as sheer nest-feathering. At least in later years, he accepted the need for a balanced economic order where manufacturing received the same sustenance as agriculture.[28] But it was the squire of Monticello who, more than any other, established the precedent that only "vendible matter" could be patented, never "a mere principle."[29] By this he meant not that machinery should be rewarded over ideas, but that tangible benefits ought accrue only to those who could produce a tangible product. Nor did he believe property rights attached to unpatented inventions, for as the dictum ran, "Land, yes, chattel, yes, but a chose in action, no."[30] The primary consideration, then, was public interest. Congress could and should provide for exclusive privilege when expedient for the common good, but Jefferson himself refused to patent his ingenious plow. Better, he thought, for the community to have free access.[31]

In Jefferson's reality, the worlds of science and public welfare were evidently one. Truth and freedom went hand in hand. Each was a marketplace, the same self-nourishing marketplace. If passing legislation would help one, it would help the other. If patenting invention was good only for the inventor, it was certainly not good for science.

George Washington aside, Benjamin Franklin was the most lionized American of his time.[32] He caused a sensation both in this country and abroad when, by flying a kite in a thunderstorm, he was able to demonstrate that lightning was not some supernatural force, but rather, was a gigantic electrical charge, similar in action to the workings of the Leyden jar. As a matter of fact, "Franklin was the most important scientist of

26. Hofstadter, *American Political Tradition*, 23.

27. Graham v. John Deere, 383 U.S. 1, 7–10 (1966).

28. Inlow, *Patent Grant*, 56.

29. A. Hunter Dupree, *Science in the Federal Government* (Cambridge: Harvard University Press, 1957), 13–14. Italics omitted.

30. Inlow, *Patent Grant*, 71.

31. These comments reflect my reading of Jefferson's sentiments and Inlow's interpretations of same. Ibid., 71–72.

32. C. Herman Pritchett, *The American Constitution* (New York: McGraw-Hill, 1959), 17.

the eighteenth century," not because of his kite exploits, but because he made the study of electricity an exact science and because he championed experimentation as the archetypal mode of scientific investigation.[33] He framed hypotheses with clarity and parsimony, he tested each empirically by manipulating sundry conductors, he quantified his data where possible, thus allowing him to compare the processes and properties of electrification, and he built a theory of electricity as a natural phenomenon, predicating his generalizations on the positive-negative measurement scheme we today take for granted.[34] Like many scientists, he was led by his experiments to the door of invention, but, like Jefferson, he declined to petition for patent rights where he might have. For him, "openness" and the sharing of knowledge as facets of the public good were sine qua nons.

In assessing the Founders' perception of science, Woodrow Wilson contended that their infatuation with the "separation of powers–checks and balances" model stemmed directly from their uncritical acceptance of Newtonian physics. In his opinion, the Philadelphia convention saw government as a machine akin to the universal order, its component parts, like the heavenly bodies, responding to immutable laws of countervailing influence.[35] Perhaps John Marshall did see himself deducing constitutional principles in the fashion of Newton deducing scientific principles; and perhaps Wilson was in part constructing a straw man so that his theory of "government as living organism" susceptible to Darwinian analysis would look better by the comparison. Whatever the final judgment, we should not be surprised to find learned men of affairs somewhat under the influence, if only in their acculturation, of generally acknowledged scientific truths.

Our concerns, though, are of a different order. How, we ask, did the Framers perceive science as a corpus of cultural and intellectual activity, the value of which they professed by according it constitutional status and the enhancement of which they believed could likely be achieved by popular support through law? What answers do we glean from the words and actions of Jefferson and Franklin, scientists and Framers both?

From the standpoint of *procedure,* these stalwarts well understood the necessity for empirical methodologies in general and experimental designs in particular. Science as abstract speculation was for them a con-

33. J. G. Crowther, *Famous American Men of Science* (Freeport, N.Y.: Books for Libraries, 1969), 151. See also I. Bernard Cohen, *Franklin and Newton* (Philadelphia: American Philosophical Society, 1956), 83.

34. Crowther, *American Men of Science,* 72–73, 77–78; Cohen, *Franklin and Newton,* 285–87, 299–301, 304, 312, 343, 350, 489.

35. Woodrow Wilson, *Constitutional Government in the United States* (New York: Columbia University Press, 1908).

tradition in terms. Franklin, indeed, was exceedingly fortunate not to suffer serious injury from his kite-flying escapade. The point is, he was willing to risk considerable personal consequences to further his research objectives. Of course, neither man ever did anything to jeopardize the well-being of others, and there is no warrant for supposing that they would have rationalized such dangers in the name of truth-seeking. Rather, given their twin commitments to science and democracy, one can readily envision them proposing legislative middle grounds to accommodate the core needs of research and social welfare. From the standpoint of *substance,* both construed science broadly, Jefferson the more so with his thinking and Franklin the more so with his doing. Neither considered science and social responsibility necessarily the winners when patents were granted on their behalf. For them, the marketplace of scientific ideas and the needs of an enlightened citizenry were too complex, too interdependent, for any all-or-nothing policy solution.

Summing up Jefferson's and Franklin's contributions: There is no field of investigation, no meritorious research question, no generally recognized mode of inquiry in science today which is either so esoteric or so culture-bound by contemporary standards that their collective wisdom is unable to comprehend it. Our notions of science are more refined, but we have built within their intellectual horizons.

Big Science—Bigger Government:
The Laws of Nature and Bureaucracy

"Americans are notoriously more interested in inventing gadgets than in studying the basic laws of nature."[36] Such has been the vigor of a prevailing middle-class social structure, steeped in the utilitarian work ethic. The following presentation argues that this infatuation with technology and its fruits has colored, to a greater extent than any other factor, scientific growth within the context of American constitutional history. Primary consideration will be devoted to the forms of scientific activity nurtured in federal law and to the bureaucratic arrangements contrived for housing these investigations as mature, long-term social commitments. Little attention will be accorded the question whether and the degree to which executive or legislative responsibility for "science policy" is the more feasible, inquiries that contemporary politicians as institutional gadgeteers have perhaps taken too seriously.[37] The impor-

36. Don K. Price, *The Scientific Estate* (Cambridge: Harvard University Press, 1965), 208.

37. Ever since President Eisenhower appointed a special assistant for science and technology, virtually every chief executive has conferred upon that personage a particular blend of roles, responsibilities, and relationships within the administrative power structure.

tant point is that we cannot understand Washington's reaction to the recombinant DNA debate unless we first examine Washington's record over time in promoting and regulating other species of science. Moreover, we cannot formulate a living constitutional profile of government-science relationships until we first (a) see how the state has influenced science and (b) compare and contrast the kinds of science influenced.

The United States could not be classified as a hotbed of scientific research in the nineteenth century. While Europe was producing the likes of Darwin, Pasteur, and Gauss, this country could claim only J. Willard Gibbs as a first-class figure. Franklin's contributions to physics had never really been appreciated here; he was remembered mostly as the inventor of the lightning rod, bifocals, and a unique stove. America's most revered "scientist" was Edison, whose greatest contribution, one authority has suggested, lay in "the reduction of invention to a rational and business-like activity."[38] And, of course, Bell, Whitney, and Howe had given us the telephone, the cotton gin, and the sewing machine.

Not surprisingly, government intersections with science at this time were few, far between, and almost entirely mission-oriented.[39] Jefferson, as secretary of state, had been a member of our first patent-granting agency during George Washington's administration, where he played an important role in establishing operational definitions for "discovery" and "invention." (These labors became so technical that Congress was constrained to institutionalize the Patent Office in 1836.) But it was President Jefferson who launched the first significant federal scientific effort. He set up the Coast Survey and instructed it to map the oceans in aid of navigation and trade. He also organized the Lewis and Clark explorations, which had the salutary effect of providing important scientific information about the Northwest frontier while, at the same time, allowing the nation to advance major territorial and commercial interests.

It is small wonder that the climate of opinion required much deliberation and persistence before Congress was able to establish the National Academy of Sciences in 1863. Commissioned to investigate scientific matters of public concern, the academy's structure and functions even

---

Adding to the confusion has been a series of executive orders and statutes creating and abolishing assorted science advisory councils within the Executive Office of the President. These maneuvers, of course, reflect the chief executive's interest in and definition of science, not to mention the concerns which Congress may at the time express regarding science as politically relevant activity.

38. Crowther, *American Men of Science,* xii.

39. The classic study of these interrelationships prior to the Second World War is Dupree, *Science in the Federal Government.*

today raise interesting constitutional questions. While the literature very often characterizes the organization as a private, nonprofit outfit—membership is restricted to the nation's foremost scientists, who replenish their own numbers—its *raison d'être,* as prescribed by law, is wedded to public purpose. That the academy could formulate whatever admissions criteria seemed in its judgment appropriate, contending, in effect, that organizational activities are not "state action" under the Fifth Amendment's due process–equal protection component, would hardly square with contemporary Supreme Court wisdom. But the larger lesson presented here is that Science, American Style, is very often a curious blend of public-private behaviors, a constitutional phenomenon of great importance for understanding the recombinant DNA debate. We may note, incidentally, that for all the struggle attendant in its founding, the academy remained virtually lifeless until World War I.

During this century, there was but one federal agency ministering to the needs of pure science for its own sake, the Smithsonian Institution, established in 1846. The Smithsonian came into being only because an anti-English Englishman willed his properties to the United States on the condition that they be used to fund "an Establishment for the increase and diffusion of knowledge among men." Predictably, the Calhounists argued that federal authorities lacked power to promote science except via the patent clause; hence, legislation approving the facility was impermissible. Eschewing niceties over federalist theory, however, Congress took charge of the estate after prolonged debate. Once off the ground, though, the institution became, in the hands of Joseph Henry and his successors, not only the seat of national museum policy but an important research center. Emphasizing that he was heading up a *private* foundation for which the government was only a trustee,[40] Henry fostered experiments in such diverse areas as atmospheric acoustics and the light-giving capacities of many oils. Later, the institution helped set up the Woods Hole laboratory for biological studies. Today, the Smithsonian sponsors extensive research based chiefly on the vast collections for which it is noted, though the many investigations conducted at the Astrophysical Observatory also come under its wing.

Much more important in the scheme of things was the burgeoning relationship between Washington and the farm community. In one fell swoop, the Congress of 1862 established the Department of Agriculture and passed both the Homestead and Morrill acts. Over the next few decades, research bureaus took root in the USDA while experiment stations developed in the land-grant universities. We have here the first permanent instance of "extramural" government research so commonplace today, wherein outside agencies and investigators perform scientific

40. Crowther, *American Men of Science,* 207.

tasks at Washington's behest.[41] Note also that the arrangement circumvented "dual federalist" constitutional theory by appropriating money for the general welfare through targeted assistance to consenting state institutions. In 1938, the Department of Agriculture could boast that it was underwriting 40 percent of all federal spending for science.[42] In 1982, however, USDA was coming under fire for failure to make the most of its basic research opportunities. While many experts believe that recombinant DNA explorations may one day provide a bountiful harvest in new and useful crops, the department apparently lags behind in this area, concentrating on the conventional aim of promoting productivity through applied investigations.[43] Still, USDA manages to devote a larger slice of its research budget (35%) to basic science than does any other cabinet-level department, save Interior.[44]

As the nation moved into the twentieth century, increased industrialization and urbanization began making their influence felt. From Carnegie and Rockefeller came charitable organizations and institutions of higher education, their avowed tasks being to encourage scientific understanding. The large firms, having reaped a harvest in the era of laissez-faire, began to establish extensive research operations, of which today the Bell Laboratories are the most famous. To these tendencies, Washington responded with the National Bureau of Standards (1901), the Public Health Service (1912), and the National Institutes of Health (1930). The NBS was created in pursuance of Congress' power to maintain criteria for weights and measures, but soon thereafter the need arose to investigate "the determination of physical constants and the properties of materials, where such data are of great importance to scientific or manufacturing interests."[45] On the government's organization chart, NBS now sits, along with the Patent Office, in the Commerce Department. Most recent evidence shows that basic research, as agency leaders

41. J. Stefan Dupré and Sanford A. Lakoff, *Science and the Nation* (Englewood Cliffs: Prentice-Hall, 1962), 5–6.

42. Harvey Brooks, *The Government of Science* (Cambridge: MIT Press, 1968), 24.

43. Nicholas Wade, "Another Look at Agricultural Research," *Science* 215 (January 29, 1982): 483. And just when the agency was attempting to redress the balance by fostering a mammalian gene-transfer program—specifically, inserting human growth hormone DNA into pigs and sheep—a coalition of environmental activists has moved to halt the experiments on grounds, essentially, that they would wreak havoc with the laws of nature. Jeffrey L. Fox, "Rifkin Takes Aim at USDA Research," *Science* 226 (October 19, 1984): 321.

44. National Science Board, *Basic Research in the Mission Agencies: Agency Perspectives on the Conduct and Support of Basic Research,* NSB-7-1 (Washington, D.C.: Government Printing Office, 1978), 6. Hereinafter referred to as *Basic Research.* Cf. National Science Foundation, *Federal Funds for Research, Development, and Other Scientific Activities,* NSF 77-301, XXV (Washington, D.C.: Government Printing Office, 1976), 16. Hereinafter cited as *Federal Funds.*

45. Quoted in Dupree, *Science in the Federal Government,* 273–74.

Table 1.1. Nobel Prizes awarded to Americans, 1901-1950

Years[a]	Number of American recipients	Percentage of all recipients by field			
		Physics	Chemistry	Medicine and physiology	All scientific fields
1901–24	4	7%	4%	4%	5%
1925–50	26	26%	20%	36%	28%

NOTE: I have coded as Americans those who were U.S. citizens at the time they received the award. Though they may have done their groundbreaking research elsewhere, they had since become members of the American intellectual community and were contributing to the growth of science here.

All percentages are rounded down.

[a]Because of wartime, the prizes were not tendered for the years 1916 and 1940–42. Each time period, then, comprises 23 years.

define the term, has fallen below 20 percent of the bureau's experimental workload.[46] The Public Health Service came into being as a result of the Progressive movement. Putting heavy emphasis on combatting communicable disease, the service, like NBS, found it necessary to branch out, with resident scientists developing shops in chemistry, bacteriology, and zoology. At first, the NIH was simply the ever more ambitious research arm of the PHS, housed in elaborate new quarters at Bethesda. Americans have generally been much more willing to invest in medical education than in most other academic endeavors, and NIH's mission to encourage research of that description has provided excellent leverage in the quest for resources. To sum up: Public support for pure science had been old hat on the Continent, but such subsidy came in via the side door here, spinning off, as has been seen, from various government enterprises designed to make life easier, safer, and more convenient for the average citizen.

How does one chart increases in national scientific accomplishment? An enlightening profile emerges when the number of Nobel Prize–winning Americans for the first quarter of the twentieth century is compared with the number of American recipients for the second quarter (Table 1.1). If we define basic science as the study of the laws of nature—a generally accepted characterization—and if we presume a high correlation between outstanding work in those sciences for which the prize is given and attaining the status of Nobel Laureate, then the data show that the number of Americans deemed worthy of recognition for significant achievement in pure science increased more than sixfold. Further,

46. *Basic Research,* 46.

Americans garnered a fivefold greater share of trophies awarded in these areas during the second period than did their predecessor countrymen, and this trend was fairly uniform over the entire spectrum of research endeavor. Science in the U.S.A. was slowly coming of age.

What brought this country to the forefront was World War II and its consequences. Because "the power to wage war," under the American constitutional system as under all other systems, "is the power to wage war successfully,"[47] the federal government entered into arrangements for scientific projects without parallel. To be sure, there were helpful precedents stemming from our experience in World War I. President Wilson had activated the National Academy of Sciences by vesting within it the National Research Council. The NRC was told to integrate government-business-university science in a manner which would best effectuate the national interest. Of course none of that had lasting effect except as part of our "constitutional memory." The central coordinator for what might be labeled "American Science against Hitler" was the Office of Scientific Research and Development (OSRD). Lodged in the Executive Office of the President by direction of Franklin Roosevelt, OSRD was instructed to concentrate on all aspects of weapons improvement from R (research) to D (development). The classic example of its efforts was the Manhattan Project. On December 2, 1942, University of Chicago scientists, operating under special pleas from the executive branch, accomplished the first contrived nuclear chain reaction. With the National Academy supplying critical advice, Roosevelt asked the Army to work up a contractual agreement with the University of California for organizing the Los Alamos National Laboratory. There, the atomic bomb was built through a joint effort featuring federal funding, industrial hardware, and university research and administration.

It was Vannevar Bush, the director of OSRD, who drew up the blueprint for postwar relations between the national government and science. In his report to Mr. Roosevelt, Bush advocated establishment of a national research foundation. Why? For one thing, new "products" such as penicillin and plastics cost money, so it was time Americans as a whole helped pay the bill. For another thing, Europe was ravaged, and the United States could no longer lean on others for scientific knowledge while it concentrated on technological applications. Moreover, Hitler had taught the nation to be eternally vigilant; research for defense was appropriate in the face of as yet unspecified enemies.[48]

47. Chief Justice Charles Evans Hughes, speaking for the Supreme Court in Home Building and Loan Association v. Blaisdell, 290 U.S. 398, 426 (1934).

48. Vannevar Bush, *Science—The Endless Frontier* (Washington, D.C.: Government Printing Office, 1945), 1-5.

18

Director Bush's call in the name of science appears today a masterful exercise. Again and again, the voice of the true educator and patron of knowledge rings out. For him, institutions of higher learning, whose faculties are protected by academic freedom, have a unique responsibility to challenge the faces of bias and habit with the harsh realities of truth. Government's research forte, he thought, is the "background" study performed so admirably by the Bureau of Standards; its domain should be extended to work of public importance covered insufficiently at the campus level. In his view, industry ought to be the home of applied investigations, for there the profit motive controls and should control. *But where pure and applied research go on under the same roof, the cheaper will drive out the dearer.*[49] Federal aid to basic science should be controlled by one agency run by scientists themselves; its task, Bush argued, would be to funnel support to the most able investigators working in the academic community. While the new foundation must be responsible to the people's representatives, "the internal control of policy, personnel, and the method and scope of the research [must be left] to the institutions themselves."[50] And all of this Bush hinged upon a set of assumptions, implicitly *constitutional* in scale: "Health, well-being, and security are proper concerns of Government, [therefore] scientific progress . . . must be . . . of vital interest to Government."[51] As the discoveries of pure science belong in the public domain, federal coffers should sponsor such research in the name of the public.[52] With respect to all the preliminary rhetoric about technological fruits, the scientific community could be more than thankful that the OSRD director knew how to package his platform and sell it in the only way American politics could process it.[53]

There is no way to investigate the intersections between the federal government and science in the post–World War II era without descending into a vast bureaucratic maze. Not the Weberian hierarchical superstructure, to be sure; rather, it is a sort of first cousin to Morton Grodzins' "marble cake" characterization of American federalism, which he described thusly: "When you slice through it you reveal an inseparable mixture of differently colored ingredients. . . . Vertical and diagonal lines almost obliterate the horizontal ones, and in some places there are unexpected whirls and an imperceptible merging of colors."[54] Today

49. Ibid., 14, 26, 77.
50. Ibid., 27.
51. Ibid., 6.
52. Ibid., 76.
53. In accord is Price, *Scientific Estate,* 2, 4.
54. Morton Grodzins, "Centralization and Decentralization in the American Federal System," in Robert A. Goldwin, ed., *A Nation of States* (Chicago: Rand McNally, 1963), 3–4.

there are approximately forty agencies in American national government engaged in "scientific activities,"[55] an array that seems to defy constitutional generalization. I will try my hand.

When peace came and OSRD went out of business, the defense establishment had no intention of waiting around until Congress approved what would be the National Science Foundation. First off the mark was the Navy Department, which received statutory authorization in 1946 to organize the Office of Naval Research. Its scientists have discovered the Van Allen radiation belt, and its monies have supported long-range acoustics studies at Columbia University's Lamont Geological Observatory.[56] But, of course, applied investigations must come first in a facility designed to address national security concerns, and, in fact, the Navy's overall level of funding for basic science has fallen recently to about a third of its total research budget.

One year after ONR's birth, the Department of Defense came into being, bringing together the traditional cabinet-level armed services agencies and taking political responsibility for military R&D. In the early 1960s, Defense received about 80 percent of all federal money allocated for science and technology,[57] yet provided less than 15 percent of the support funneled from the federal treasury to the campus.[58] Those recipients currently include—under a broad but defensible definition of "campus" —government-owned but university-operated "think tanks" maintained by MIT (Lincoln Labs) and the California Institute of Technology (the Jet Propulsion Laboratory). There, DOD's expenditures go largely toward the applied science one associates with space vehicular research projects. DOD lists approximately 18 percent of research monies as earmarked for pure scholarship, with about 38 percent of these going to university endeavors. This work, like the science subsidized at the national laboratory think centers, is entirely of the contract variety, whereby an agency solicits outside brain power to perform certain tasks according to a set of quid pro quos. Certainly the power to make contracts with Harvard would seem to be at least as broad as the power to make grants-in-aid to the University of Illinois for experiments in crop genetics. Interestingly, the Pentagon reports that virtually none of its on-campus research operations are of the "applied" species,[59] seeming to contend that the heavy investments by the Defense Advanced Research Projects Agency in computer science at MIT and Stanford are best conceptualized

55. For an appropriate definition of "scientific activity," see National Science Foundation, *Federal Organization for Scientific Activities, 1962,* NSF 62-37 (Washington, D.C.: Government Printing Office, 1963), 575.

56. Brooks, *Government of Science,* 110–25.

57. Ibid., 3.

58. Price, *Scientific Estate,* 41–42.

59. The data for current Defense Department science are in *Federal Funds,* 14.

as basic explorations. Nonetheless, because the funding agent is concerned primarily with technological consequences, because the means of support is in the form of a contract (where individual initiatives are specifically reserved) rather than a grant, because the work entails substantive questions of greatest relevance to engineering as a discipline, and because much of this research is devoted to artificial intelligence, considerable aspects of which can hardly be considered nonutilitarian or abstract, on balance it is much more accurate to place these studies in a twilight zone category between pure and applied investigations.

The atomic bomb, of course, has always posed special problems. Relying upon the constitutional principle that "the power to wage war includes the power to prevent it,"[60] Congress vested in the federal government a complete monopoly on research and technology involving fissionable materials. Hence, patent applications in this unique field of inquiry are not, and never have been, honored. Since 1977, the Department of Energy has exercised total control over nuclear science through a variety of bureaucratic arrangements that would tax the credulity of any management systems expert. We at least know that almost all work is developed extramurally under the contractual agreement format. As for type of research performed, Sandia's agenda exemplifies applied investigations specialization, Brookhaven's agenda exemplifies pure research commitment, and Oak Ridge's agenda is of a soup-to-nuts variety. There is also considerable contractual work done within educational and industrial establishments themselves; as in the case of DOD, these connections should be viewed in contradistinction to the tasks performed at most of the federally financed research centers.[61]

In the wake of the Russian Sputnik triumph, Congress created the only federal agency currently devoted to the concerns of a single scientific-technological field, NASA. Exploration of outer space, of course, is its stock-in-trade, and TV watchers are well familiar with Cal Tech's Jet Propulsion Laboratory, at which exotic technological effort is mounted in the name of astrophysics and planetary studies.[62] JPL's research image does not become more distinct when we find that 30 percent of its 1983 research budget was funded by the Defense Department, as NASA

60. Edward S. Corwin, *The Constitution and What it Means Today,* 11th ed. (Princeton: Princeton University Press, 1954), 68.

61. Brooks, *Government of Science,* 135. I share Brooks's opinion (p. 186) that the Cambridge Electron Accelerator and the Lawrence Radiation Laboratory at Berkeley, though government supported, are essentially bastions of pure research and part of the university culture.

62. If a generic label is to be applied, the Jet Propulsion facility is perhaps most accurately called an "academic think tank–do tank."

buffets under the weight of severe fiscal constraints.[63] Both industry and higher education have always had a big stake in the space program, with the former specializing in how-to aspects while the latter focuses in on the esoteric cosmological questions.

In 1950, Congress at long last gave birth to the National Science Foundation. True to Vannevar Bush's conception, NSF awards *grants* (not contracts), largely for basic research, to institutions—usually those of higher education—in the name of principal investigators who have tendered unsolicited proposals. The grant procedure, which the NIH had already employed successfully, is considered far superior to the contractual arrangement in the context of pure science because it can be an exercise in futility to estimate legal obligations where goals and directions are often unpredictable. Contracts are feasible when government purchases services, but grants are more appropriate when government sponsors projects to expand the known pool of knowledge.

This system of grantsmanship is implemented by the process of peer review. That is, NSF is organized around clusters of scientific disciplines, one of these being labeled Biological, Behavioral, and Social Sciences. A Division of Physiology, Cellular and Molecular Biology is located in this grouping. The several programs within each division have at their disposal panels of distinguished specialists who are asked to consider the merits of potential projects on the basis of such criteria as proposal competence, intrinsic worth, and relevance to pressing social problems or technological needs. The relevance guideline comes into play particularly on those occasions when applied research grants are under consideration.

How are applications featuring recombinant DNA research processed? The typical proposal is referred to the Genetic Biology Program, which, according to one informed estimate, received more than 300 such requests, or 90 percent of its business, in 1981. Peer review takes two forms. First, NSF forwards application papers to six "outside" experts; second, two members of the Program Subcommittee, a peer group whose members serve set terms, also work up evaluations. So while the ground rules mandate three in-depth analyses, each bid normally is accorded eight reviews. Final recommendations are the responsibility of the program director for Genetic Biology. This full-time NSF staff officer files opinions on all applications, basing judgments not so much on peer review ratings as on the nature of peer review criticisms and accolades. The best estimates are that from 15 to 20 percent of proposals are funded, but the number would rise to between 40 and 45 percent if recipients did not

63. M. Mitchell Waldrop, "Jet Propulsion Lab Director to Resign," *Science* 216 (April 16, 1982): 276.

22

obtain federal assistance elsewhere. NSF and NIH cannot provide joint support.

From the program level, favorable decisions are routed to a section head—the crush of business has lately caused NSF to provide subsets of programs with "section oversight"—and then to the director of the Physiology, Cellular and Molecular Biology Division. On rare occasions, one or the other official will overrule the program director's justifications. Special cases and appeals crop up, of course, and here the assistant director for the Biological, Behavioral, and Social Sciences Directorate would probably get involved.[64]

Vannevar Bush had assumed that democratic values and scientific integrity could each be served adequately should the president appoint, with Senate approval, the NSF governing board of research experts. But some scientists thought a director, named by the chief executive and reporting directly to him, ought to sit atop the pyramid of "disinterested" scholars. In that way, popular control might check clientele special interest. This tension, which anticipates aspects of the recombinant DNA debate, was resolved when Congress provided for a director and a National Science Board, the former carrying the ball with respect to NSF "politics," the latter sitting as overall policy framer. The board also assumes final responsibility for grant approval, but in practice, it looks at specific proposals only where a very large sum is being awarded for a lengthy period of time, most infrequent circumstances. NSF is essentially an independent agency, in this respect being akin to NASA, but unlike the space program it is not organized around a single scientific endeavor with clear technological consequences. Hence, its political leverage is much weaker.

A second bone of contention in setting up the National Science Foundation involved the thorny matter of patent rights. Some scientists, Bush among them, thought patents should issue to NSF-sponsored inventors and institutions unless "special cases" warranted a contrary determination.[65] Others would have put an embargo on claims of exclusivity, hoping thereby to discourage applied, commercially feasible science from feeding at the public trough. In another compromise, it was decided that patents might be granted, but that NSF would be obligated to employ them only as inducements for pure research. In fiscal 1980, the foundation evaluated seventy requests for patent application clearance and ruled favorably on sixty-one of these.[66] The others now lie in the public domain because of publication.

64. Interview with Delill Nasser, Acting Program Director, Genetic Biology Program, NSF, February 8, 1982.

65. Bush, *Science,* 32.

66. National Science Foundation, *Annual Report 1980,* 126.

From the foregoing remarks, one would surely infer that the country's scientific leadership intended NSF to take the initiative in encouraging genetic investigations as a whole and recombinant DNA innovations in particular. Obviously, such has not been the case, and as a matter of fact, in 1979 the foundation provided only twenty-one cents of the over-all federal basic research dollar.[67] While policymakers were debating the structure and function of NSF in the late 1940s, the National Institutes of Health came into existence in their present supervisory form and commenced a major effort in the medical sciences research arena. For some time now, NIH has been the largest single contributor to pure research at the university level. What organizational dynamics feature these operations?

NIH has always been attached to the Public Health Service, which, since 1953, has itself been lodged in the Department of Health and Human Services (formerly HEW). The department is under congressional mandate to conduct investigations on all health-related matters. To this end, PHS has an overriding concern for the many life sciences, and NIH, as chief overseer of health research, has converted that commission into a 67 percent responsibility for federal biomedical R&D. Such is NIH's importance that its director, who during the heat of the recombinant DNA debate was Donald Frederickson, no longer reports and provides technical expertise to the surgeon general but, rather, to the HHS assistant secretary for Health and Scientific Affairs. The eleven institutes themselves are generally organized around the investigation and treatment of particular diseases. Among these, the National Cancer Institute (NCI) and the National Institute of Allergy and Infectious Diseases (NIAID) are most important for our purposes, but there is also a National Institute of General Medical Sciences (NIGMS), which supports basic biomedical research having no especial relevance to the work of the more mission-oriented shops. NIH, in sum, is a very complicated place with a wide variety of political structures and involvements. For example, there are both extensive laboratory and clinical facilities at Bethesda; grants include not only awards for research but money for regional medical programs, state control programs, and community demonstration projects; and latest figures show that the National Cancer Institute is supplementing its generous support for scholarly endeavors with almost 2,000 research contracts valued at $268 million.[68] Health science is big science, big business, and big government.

Peer review has been the order of the day at NIH since 1946; and in 1974, Congress went so far as to insist that the agency employ the pro-

67. National Science Board, *Science Indicators—1978,* NSB-79-1 (Washington, D.C.: Government Printing Office, 1979), 70.
68. *1980 NIH Almanac,* NIH Pub. No. 80–5, 154.

cedure in assessing all applications for grants and contracts. The modus operandi works more or less in the fashion of NSF decisionmaking. There are seven review sections, each of which includes a battery of study sections or review groups, organized, in this case, not by disease but by academic field. For instance, the Biological Sciences Review Section comprises eleven subunits, covering such specialties as genetics, molecular biology, and microbial genetics. Each of these review groups is manned by peers from across the country and makes recommendations, first, as to approval or disapproval and, second, providing numerical scores for every proposal deemed worthy of endorsement. On those infrequent occasions when expertise is lacking in the study section, outside advice will be solicited, but ordinarily the group relies on detailed writeups prepared by the two panelists assigned to each request. Review group executive secretaries sit in on all discussions and votes; moreover, only they have access to the confidential quantitative scores. They prepare "pink sheets" for all applications, containing the pertinent pros and cons. While about 5 percent of projects are considered unworthy even of numerical evaluation, even these will be forwarded "upstairs."

Applications go from the initial review group (IRG) to the advisory councils for each institute. These councils, made up of other peers plus lay representatives possessing special skills or interests in an institute's mission, render final determinations as to both acceptance and degree of funding. Again, grants are awarded to institutions in the name of investigators.

Tracking a hypothetical recombinant DNA proposal through NIH turnstiles is impossible, because Bethesda keeps no special records for cloning projects that are exempt under prevailing guidelines. Nor does it even require campus authorities to maintain such data, though, as I point out in Chapter 4, many of them do. So information is limited to those experiments—allegedly 10 percent of the total—which are considered provocative enough that NIH constraints are deemed obligatory.[69] This sample is clearly a biased one, and even then Bethesda does not seem equipped to provide a complete picture. Still, I present what is available.

During the October 1981–May 1982 review cycle, NIH received 704 grant applications for extramural support of nonexempt gene-splicing endeavors. The vast majority (85%) were assigned either to the Biological Sciences or the Biomedical Sciences Study Sections, with the former maintaining a slightly larger share of the workload. As might have been anticipated, the most active among these review groups were Experimen-

69. Letter received from William J. Gartland, Jr., Director, Office of Recombinant DNA Activities, NIH, March 15, 1982. The data which follow are culled from a computer printout provided by Asher A. Hyatt, Chief, Biomedical Sciences Section, NIH.

tal Virology (97) and Virology (90), for research addressing pathogenic viruses is obviously highly sensitive. Following close behind were Molecular Biology (88) and Genetics (73). These four study panels handled about half of the application pool, with the rest being parceled out to the other 65-odd sections.

To which NIH bureau, institute, or division (BID) did these projects go following initial evaluation? Taking jurisdiction over the lion's share were NIGMS (205), NCI (195), and NIAID (115). Thus, 73 percent were deposited with three of the eleven advisory councils. The typical cloning entry in this select batch of 704 is one which was judged, first, in the Experimental Virology Study Section and was then sent to the National Cancer Institute for possible final acceptance and funding. Next most common is a proposal that commenced in Genetics and was then dispatched to the General Medical Sciences Institute. There were 66 of the former and 58 of the latter. All told, 255 grant applications, or 36 percent, flowed from the four most active review groups to the three most active BIDs. NIH officials were asked to supply data regarding acceptance rates and dollar support for recombinant DNA projects, but they declined to provide this information.

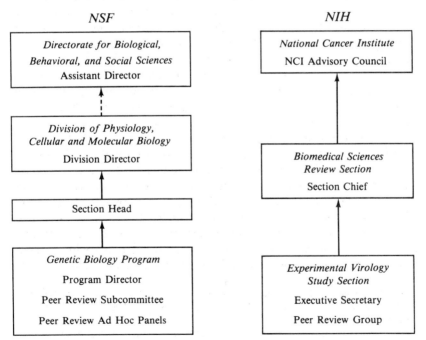

Figure 1.1. Organizational Review Flow Chart for Recombinant DNA Research

26

A comparative bureaucratic profile of NIH-NSF funding procedures for the typical gene-splicing research investigation discussed above takes the form diagrammed in Figure 1.1. Not only does the NIH regimen, with its two-tiered peer review structure, appear somewhat more elaborate and formalistic from the vantage point of the recombinant DNA investigator, but also one must remember that there are eleven institutes and only six directorates. NIH's greater organizational "density" is in part a function of its assorted mission interests, but it is also a function of the fact that the agency has much more money to dispense and can therefore tender many more awards, as the data for FY 1980 show: NIH awarded 18,500 research grants at a cost of $2 billion, while NSF awarded 14,100 grants at a cost of $781 million.[70] In terms of support for cloning studies, NSF, which keeps complete records, reports 121 grants at a cost of $9,830,644.[71] We can only guess at the comparable NIH numbers, but even if we assume a minimum acceptance estimate of 15 percent (bearing in mind that NSF funds 15–20% of recombinant DNA proposals), then Bethesda would be subsidizing 106 *nonexempt* research projects alone.

As the National Institutes of Health provide financial assistance to the great majority of tax-supported gene-splicing exercises, and as the courts have found that recombinant life forms fall within the language of the patent statute, NIH's policy regarding an inventor's exclusive rights is of considerable relevance. The Department of Health and Human Services has developed two instruments for implementing this policy. On the one hand a university may sign an institutional patent agreement (IPA) which affords that facility a first option to own "vendible matter" produced with HHS funds, so long as minimal strings are acceptable. The most important of these conditions vest in the government a royalty-free license to employ the artifact, place a limit on the institution's right to arrange exclusive contracting, insist that the public be informed of said discoveries once patenting has been consummated, and secure reversionary rights to the taxpayer should an institution fail to commercialize the invention. HHS, at last count, has worked up seventy-two IPAs, and NSF, which also employs the device, has negotiated twenty-nine. A second mechanism allows research units to approach HHS on a case-by-case basis; in these instances the department awards patent rights to petitioners 90 percent of the time under the terms described above. So, as

70. These figures, which are approximations, are taken from *National Institutes of Health—Research Grants,* NIH Pub. No. 81-1042, 1, 3; National Science Foundation, *Annual Report 1980,* 1, 21, 49, 69, 87, 105.

71. Letter received from Antonie Blackler, Director, Division of Physiology, Cellular and Molecular Biology, NSF, February 26, 1982.

matters stand, academic centers are encouraged to go shopping for prospective purchasers of exclusive and nonexclusive rights, reaping the rewards of royalty payments should their searches be successful. Of course, it is up to the college or university to work out whatever remunerative arrangements it desires with "employee inventors," viz., faculty. It is extremely significant that under the prevailing rules, anyone can use the patented merchandise for research purposes; monopoly interests are abridged only when an artifact is exploited for commercial gain. In December of 1980, Congress rewrote the legislation affecting government-university patent agreements, but the new law includes basically the provisions which had been developed over the years through administrative practice.[72]

In 1977, federal officials debated whether cloning inventions should be covered under the usual rules and regulations.[73] Among cabinet-level personnel, only Justice Department spokesmen thought that heightened public interest in this area required special handling. They argued that the U.S. Government should be awarded patent rights for all federally funded recombinant DNA inventions and should dedicate these to the public. This contention was rejected because the IPA relationship had worked well with respect to vaccines for both rubella and rabies (among many other treatments), drugs, and technologies of medical consequence, providing not only a smooth transfer of knowledge from laboratory to marketplace but also a wide distribution of information through publication to interested parties. In 1978, NIH director Donald Frederickson concluded that recombinant DNA patent proposals should be processed via conventional means, adding only the stipulation that all licensees must abide by the NIH guidelines which funding recipients themselves were pledged to honor. As to how colleges and universities could enforce those terms with commercial outfits that had purchased licensing prerogatives, all parties realized that NIH faced the same compliance problems vis-à-vis its academic clients, and, absent legislation, they simply hoped for goodwill on all sides. At this writing, Stanford and the University of California are the only academic patent-holders in the recombinant DNA field, and they have issued nonexclusive licenses, for rather generous royalty terms, to such enterprises as Eli Lilly, Genentech, and Cetus. Given current conditions, these institutions could collect many millions before their monopoly expires, but that may well depend on whether the courts construe the ground-breaking Boyer-Cohen gene-

72. 94 *Stat.* 3015; PL 96–517, chap. 38 (1980).

73. U.S., Department of Health, Education, and Welfare, National Institutes of Health, *Recombinant DNA Research: Documents Relating to "NIH Guidelines for Research Involving Recombinant DNA Molecules,"* 2 (March 1978): 5–18.

splicing processes and the range of artifacts derived therefrom broadly or narrowly, if called upon, as undoubtedly they will be, to render an opinion.[74]

Figure 1.2 presents a cross-sectional representation of federal government–science organizational relationships as they stand today. Some explanatory notes are in order. Institutional structures are arrayed in a vertical sequence that is intended to reflect implicit understandings about the American constitutional system. The most basic distinction is between intramural and extramural research. While no one doubts the propriety of federal agencies employing scientific knowledge to further their own missions, and while no one doubts the authority of these agencies to place unlimited restrictions on the ways in which scientific investigations are conducted "in-house," profound legal questions inevitably arise when Washington enters into arrangements with other salient power centers. The least provocative of these from a *constitutional* standpoint seems to comprise the so-called military-industrial complex; next in line are relationships with state-owned or state-chartered entities, quid pro quos which invite analysis under such competing theories of divided sovereignty as Franklin Roosevelt's "cooperative federalism" and Ronald Reagan's "new federalism"; finally, and most fraught with tension, we have Washington's ties with educational entities, society's engine in the quest for truth and the polar opposite of government, society's engine in the quest for power.[75] Arrayed in horizontal order are the forms of scientific activity which these agreements foster, running from those government bodies most involved in applied investigations to those most involved in pure research. Here, social scientists must rely to a great extent on agency definitions of their own research undertakings, and there are unavoidable risks in so doing. It is very likely that the U.S. Geological Survey does not expend 80 percent of its investigations on pure science, as the NSF and this book define that concept.[76] And certainly the Smithsonian, which considers itself "the nation's environmental bureau of standards,"[77] cannot rank ahead of NSF itself on the basic research ladder, even when it reports a 100 percent budgeting expenditure on basic research.[78] Note that recombinant DNA efforts are largely

74. Barnaby J. Feder, "Gene-Splicing Licensed," *New York Times,* August 19, 1981, D1–D2. See also Colin Norman, "Cohen-Boyer Patent Finally Issued," *Science* 225 (September 14, 1984): 1134. For further discussion of the Boyer-Cohen technology, see Chapter 2 below.

75. Price, *Scientific Estate,* 191.

76. NSF is well aware of these problems. See *Basic Research,* xx.

77. Ibid., 226.

78. The quantitative determinants for pure and applied research funding at the "government agency" level (see Fig. 1.2) were computed from data presented in *Federal Funds,*

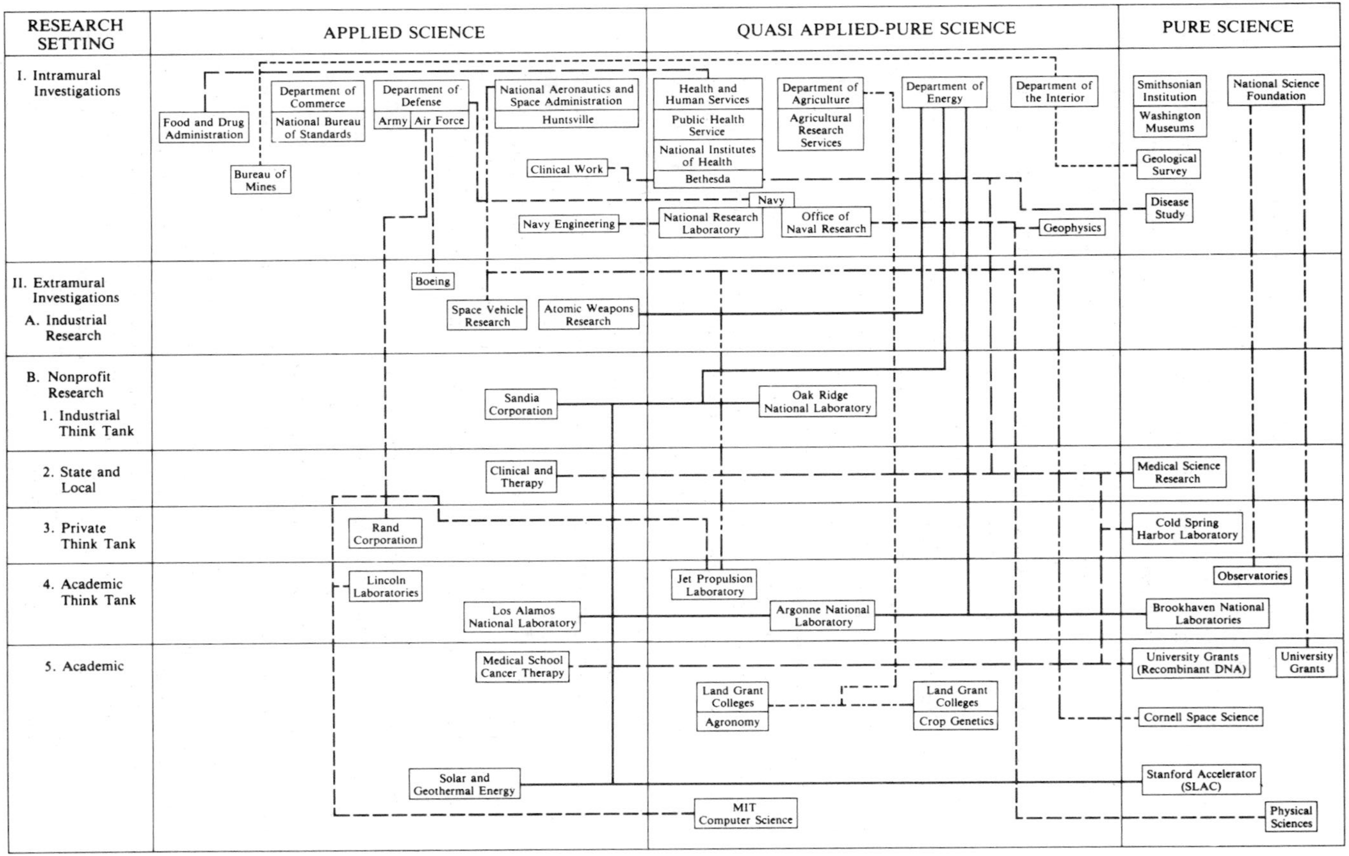

Figure 1.2. Federal Agency Funding for Classes and Locales of Scientific Inquiry

the domain of pure science inquiry, no matter how that term is employed. Finally, I should add that these depictions look very much the way they would have appeared had I charted them twenty years earlier. A single exception involves the defense establishment, which, through the impress of the 1970 Mansfield Amendment, is no longer permitted to perform research tasks unless these are directly related to departmental goals; but even this proviso today carries more bark than bite.[79]

An important question is the extent to which these ties between government and the private sector convey patterns of constitutional dimension. Two assertions are virtually self-demonstrable. First, American national government is seriously concerned about all aspects of scientific endeavor because it has learned that science is too important to suffer benign neglect. Second, in that regard, Washington has a clear obligation to orchestrate the applied research commissions necessary and proper for executing the people's business and to fund, through the project grant system, basic research on the campus, at least in the hard sciences. We are talking at this level of ordered relationships pertaining to public issues of fundamental proportion which have acquired a sense of "rightness" or legitimacy about them in the marketplace of yesterday's and today's opinion. That is, no knowledgeable observer would today call for the abolition of the National Science Foundation or the National Institutes of Health; no knowledgeable observer would today argue that these bodies should not subsidize basic investigations into the mainsprings of molecular biology; no knowledgeable observer would today argue that these research exercises should be performed under contract rather than via grants to demonstrably qualified scholars. These norms and the values they represent have become part of the Living Constitution.[80]

What has *not* achieved the legitimacy we associate with constitutional tissue is the seeming disorderliness, the labyrinthian compartmentalization evinced by many of these arrangements. Also prone to criticism is the amalgam of federal gifts and regulations which could erode, by joint carrot-stick inducement, the integrity of institutions generally considered

14–19. I have assigned governmental structures to the quasi-pure, quasi-applied sector when at least 30% of their scientific activities fall into both categories.

79. Skeptical insiders have been known to argue that the Mansfield Amendment never really had much impact, that investigators simply repackaged their project designs to demonstrate national security value. And see also the text at note 59 above, showing that Defense is no longer bashful (if it ever was) about parading its contributions to pure science. Of course, there is no reason why basic research cannot be "directly related to departmental goals," a fact which Mansfield hard-liners seemed often to overlook.

80. See generally Carmen, *Power and Balance;* cf. Karl N. Llewellyn, "The Constitution as an Institution," *Columbia Law Review* 34 (1934): 1–40.

vital to our system of countervailing authorities. Every now and again, therefore, the clarion call goes up not merely for greater centralization, but for a genuine reformulation of the government-university science relationship, a "new order" which would allow each to contribute its fair share to the public interest free from compromise over principle. This chapter concludes by examining the constitutional ramifications of that call.

A cabinet-level science department, which might well enhance both goals, has been talked about for years. Advocates contend that this agency could develop and coordinate a national science policy, something we can hardly envision given today's pluralistic, indeed diffuse, research networks. Actually, the NSF was initially charged with performing such master-plan responsibilities, but this hope foundered because its power is university-based rather than being based in the executive branch. That reality provides focus for a second supporting argument. Centralization of scientific opportunities and contributions would take considerable pressure off the academic community to play government handmaiden. Why wouldn't it be better for everyone, then, if the people had their own version of the Bell Laboratories?

The short answer to the last question is that, *given prevailing norms,* there isn't anything wrong with Washington's developing a central laboratory for basic research. Certainly Harvard's molecular biologists would not run to a Department of Science any faster than they are running to Genentech. The problem is that we would just be drawing another box on the organization chart, because, as is contended below, there is little chance our new bureau could formulate or implement successfully a coherent, viable national science policy. So the central quandary, if quandary it be, is that the prevailing bureaucratic norm structure reflects deep-seated political and scientific values—aspects of our "constitutional environment"[81]—which refuse to budge in the wake of central administration's best-laid plans. What are these values?

It is the nature of scientific inquiry that "good science" comes chiefly from good minds laboring, singly or collectively, over problems that present intrinsically fascinating characteristics for those particular investigators.[82] Wilson and Penzias did not discover remnant electromagnetic radiation, and thus apparently validate the "big bang" theory of galactic creation, because some administrator at Bell Laboratories gave them marching orders to scan the heavens and unlock nature's secrets in the

81. This concept is developed more fully in Carmen, *Power and Balance,* 26, where the term is employed to include the core time-space-cultural context which has shaped the basic institutional norms of the body politic.

82. See generally James D. Watson, *The Double Helix* (New York: Atheneum, 1968).

name of national AT&T policy. Experimentation employing recombinant DNA does not owe its birth to NIH or NSF "policy," though, of course, once shown to be rewarding, it has received nourishment at the taxpayers' expense. If genetic researchers had had to wait for government policymakers to baptize their investigations with an imprimatur before they contrived mutant life forms, then cloning, as we use the term today, might still be the preserve only of science fiction writers. Groundbreaking research, then, goes hand in hand with the diffusion-of-power principle. Or, to put the matter somewhat differently, unless the state controls basic research—and what responsible onlooker believes that it should or could?—how can the state prescribe a meaningful national research blueprint?

As to the root political values of concern here, it is hard to sell science in the United States without selling technology, invention, or national security. The federal government, plagued by a decentralized structure that makes it easy pickings for lobbyists and clientele interests, is really the last social institution in this country that can afford to ignore political reality by stripping basic science from its moorings and permitting it to wander alone through the congressional committee jungle. So to the extent that the state should hitch its pure research wagon to the star of intramuralism, that effort had better be lodged within the established, goal-oriented superstructure. And to the extent that a "Gresham's law of research" renders such poolings ill-advised, as Vannevar Bush thought they were, then all the more reason to applaud university-housed basic science arrangements.

Indeed, NIH and NSF have good track records precisely because they have tended to decentralize well. That is, not only have they blessed the scientific community with financial support, but also they have vested in that community a capacity to check and balance the Washington administrative apparatus. To be sure, peer review is hardly a synonym for science policy. But that is to be commended, because private groups, whatever their source of wisdom, generally lack public consent to mandate the ends of statecraft. The important consideration is that scientists, no matter how episodic and ad hoc their assessments, ought to play a key role in deciding who gets NIH-NSF money simply because the value judgments involved are inherently apolitical. Again, we are talking of pure research, investigations transcending specific government goals ordained by law. Some investigators may bemoan the fact that the military is prevented by congressional edict from spreading its research wings as once was its preference, and certainly many may question whether a theory of "Vietnamese stigma" should prevail, in the long run, as the guiding principle for determining the level of basic research in mission agencies, but it is doubtful whether the scientific community would shed a tear if the nation were not confronted at this time with high budget deficits and President Reagan's spending priorities. To repeat: in the context of pure investigations, the state acts most prudently when consensus among specialists, whether firm or fragile, has been achieved.

It is when these national bodies have tried to centralize through guideline and constraint that their good offices become suspect. But regulations need not emanate from the nation's capital; they can arise from local pressures and departmental meetings as well. So whether constitutional arrangements be hierarchical or pluralistic, the scientist who must have his freedom seems to have no guarantee of freedom. Do scientists possess *any* rights under our political system? That question, not the matter of bureaucratic reorganization, is the truly relevant constitutional issue of our day, and it is the issue immediately before us.

2 Cloning as American Constitutional Freedom

Civil liberty is an expanding notion in contemporary Supreme Court jurisprudence. Twenty-five years ago, the justices talked only of free speech, free press, right to counsel, and the protection against cruel and unusual punishment, guarantees specifically enumerated in the Constitution.[1] Today we read about rights of travel, association, and privacy, basic liberties emerging from the confluence of explicitly stated freedoms and worthy as well of membership among the "preferred" personal values which the judicial branch is pledged to defend.[2]

The Supreme Court has never decided a case involving the right of scientists to do research. As late as 1960, few would have expressed surprise at that fact. But in this more "activist" era, the Court has been petitioned to rule on whether Missouri could ban the use of saline amniocentesis for abortions beyond the initial trimester. And it has also been asked to determine whether New York could vest in physicians alone the power to distribute contraceptive devices to juveniles younger than sixteen years of age. In both controversies, the alleged libertarian interest was sustained, and the statutory provision at issue was declared unenforceable.[3] Why, then, would not an investigator who felt aggrieved by one or another legal obligation seek judicial redress? Is it possible that there are no scientists who feel unduly constrained?

1. Even the watershed Meyer v. Nebraska, 262 U.S. 390 (1923), was framed within the four corners of due process analysis, though the substantive gloss which the Court placed upon Fourteenth Amendment language in that decision has had significant repercussions for the development of "penumbra rights" theory. See Ira H. Carmen, *Power and Balance* (New York: Harcourt Brace Jovanovich, 1978), 272–76.

2. For the notion of countermajoritarian principles as "preferred" constitutional guarantees, see United States v. Carolene Products Co., 304 U.S. 144, 152, n.4 (1938).

3. Planned Parenthood v. Danforth, 428 U.S. 52 (1976); Carey v. Population Services, 431 U.S. 678 (1977).

This chapter builds a theory of scientific endeavor as protected constitutional liberty.[4] It examines recombinant DNA investigations as a particular form of scientific activity, reaching the judgment that cloning is entitled to some measure of constitutional solicitude. In conclusion, the state's role as research funding agent is scrutinized under prevailing judicial doctrine. I shall assert that just as government lacks carte blanche to manipulate free expression through the purse strings, so it also lacks carte blanche to manipulate scientific experimentation in general and recombinant DNA experimentation in particular by attaching capricious constraint to munificent reward. In addressing these propositions, my commentary will focus on relevant parameters of constitutional right which the Supreme Court has articulated, the attributes of science which qualify as protected freedom, and the manner in which the state permissibly can balance through appropriate regulations individual prerogatives in the laboratory and in analogous environments.

Scientific Investigation and the First Amendment

If there is one aspect of scientific inquiry which more than any other would seem to be protected speech under the Constitution, that aspect is the right of investigators to publish or otherwise disseminate their research findings. Any conclusion to the contrary must rest on the thesis that the First Amendment covers some kinds of expression but not others and that scientific discourse is among the excluded subject matters. In fact, a few notable constitutional scholars have endorsed this view, their position being essentially that the purpose of free speech is to promote a robust exchange of *political* ideas.[5] However, the predominant opinion conceives protected expression as a methodology designed to "advance knowledge and discover the truth,"[6] and most distinguished critics who have discussed the matter specifically conclude that scientific speech merits First Amendment status.[7] Moreover, there are Supreme Court dicta which, taken together, strongly fortify this consensus.[8] If there is

4. Some of the arguments presented here are embellishments and additions to the thesis expounded in Ira H. Carmen, "The Constitution in the Laboratory: Recombinant DNA Research as 'Free Expression,'" *Journal of Politics* 43 (1981): 737–62.

5. Robert Bork, "Neutral Principles and Some First Amendment Problems," *Indiana Law Journal* 47 (1971): 1–35. Alexander Meiklejohn also supported this position at one time. See *Free Speech and Its Relation to Self-Government* (New York: Harper, 1948).

6. Richard Delgado and David R. Millen, "God, Galileo, and Government: Toward Constitutional Protection for Scientific Inquiry," *Washington Law Review* 53 (1978): 363.

7. Zechariah Chafee, Jr., *Freedom of Speech* (New York: Harcourt, Brace and Howe, 1920), 370; Thomas I. Emerson, "Toward a General Theory of the First Amendment," *Yale Law Journal* 72 (1963): 881–84.

8. Roth v. United States, 354 U.S. 476, 487 (1957); Miller v. California, 413 U.S. 15, 24

expression warranting total legislative suppression,[9] those forms of "perverse speech" would, under current Supreme Court holdings, include only the "obscene, . . . the libelous, and . . . 'fighting' words." They are exceptions because their contribution to the marketplace of ideas is "of such slight social value"[10] as to be manifestly outweighed by the public interest in proscribing sundry noxious consequences. With respect to obscenity, the antisocial element which the state may reasonably suppress is the patently offensive sexual representation that accomplishes little more than a titillation of prurient appetites;[11] in the case of libel, the political process need not accord any legal protection to character assassination;[12] and regarding face-to-face insults, the mores frown upon personal invective that will most likely inspire not discussion but physical reprisal.[13] That scientific speech has redeeming social importance and that it poses no inherent threat to either individual integrity or community standards are seemingly incontestable propositions. Indeed, we might well ask why it is more important to know about Aristotle's or Hobbes's political theories than their scientific theories, especially given the fact that one cannot understand the former without comprehending the latter.

The Supreme Court has held, of course, that freedom of speech is not an absolute right even as regards content ordinarily the recipient of constitutional shelter. Urging soldiers to lay down their arms during wartime,[14] inciting to riot as a means of achieving civil rights,[15] disseminating "adult" materials to children,[16] and employing political tracts as a conspiratorial strategy to foment the overthrow of government,[17] not to

---

(1973). And see more generally Thornhill v. Alabama, 310 U.S. 88, 101–2 (1940), and Kingsley Pictures v. Regents, 360 U.S. 684, 688 (1959).

9. On occasion the justices have attempted to rationalize legislative discretion in regulating these utterances by asserting that they are not speech in any First Amendment sense. One such instance is Roth v. United States, 354 U.S. 476 (1957); another is Cantwell v. Connecticut, 310 U.S. 296 (1940). The hope is that a speech-nonspeech dichotomy will avoid the travails of balancing public needs against private rights. As this section will assert, however, such rigid distinctions do not suffice. Indeed, the Court, in its more candid, reflective moments, tends to eschew them.

10. Chaplinsky v. New Hampshire, 315 U.S. 568, 572 (1942).

11. Miller v. California, 413 U.S. 15 (1973); Paris Adult Theatre v. Slaton, 413 U.S. 49 (1973).

12. New York Times Co. v. Sullivan, 376 U.S. 255 (1964); Gertz v. Robert Welch, Inc., 418 U.S. 323 (1974).

13. Chaplinsky v. New Hampshire, 315 U.S. 568 (1942).

14. Schenck v. United States, 249 U.S. 47 (1919).

15. Feiner v. New York, 340 U.S. 315 (1951).

16. Ginsberg v. New York, 390 U.S. 629 (1968).

17. Dennis v. United States, 341 U.S. 494 (1951).

mention inspiring revolutionary conduct that poses a clear and present danger to insurrection[18]—all these the state may properly forbid. It is conceivable—though actual instances are unknown to our law—that scientific expression could run afoul of such strictures. Certainly researchers have no greater right than the average person to disseminate their ideas in a manner violative of these doctrines.

But, as we know, scientists do far more than spread opinion. Their craft may be considered a process which includes, at the least, cognition, observation, discovery, and those collegial bonds—namely, the teacher-student and teacher-teacher relationships—which form the basis of intellectual cross-fertilization.[19] The Supreme Court has had much to say about the constitutional status of interpersonal communication in the educational setting. Instruction and learning, it has found, are the cornerstones of academic freedom; moreover, the "right of association" is central to a free society and inherent in the Bill of Rights, precisely because it is an indispensable vehicle for like-minded people to exchange views and search out the truth. The justices' theme is that freedom of speech amounts to little without meaningful idea marketplaces, and that consequently, interpersonal relationships devoted to establishing and enriching those marketplaces, of which the academic environment is a salient example, require First Amendment shelter.[20] While none of these remarks pertain specifically to scientists and what they do, it is incomprehensible that classroom associations should receive a higher degree of per se constitutional protection than laboratory associations. Indeed, from the collegial standpoint, the laboratory is, in many ways, nothing but a classroom.

With respect to mentation and its corollary faculties, on the other hand, the Court has talked only in the vaguest terms. Hence, we find Justice Douglas declaring: "The state may not, consistently with the spirit of the First Amendment, contract the spectrum of available knowledge. . . . The right of freedom of speech . . . includes . . . freedom of inquiry."[21] Specifically, people do have a right to receive materials containing protected speech through the mails,[22] and they also have a right to read anything they wish—even illicit, hard-core pornography—in the privacy of their own homes.[23] If scientific speech is free expression, then the role

18. Brandenburg v. Ohio, 395 U.S. 444 (1969).
19. Delgado and Millen, "God, Galileo, and Government," 363.
20. Meyer v. Nebraska, 262 U.S. 390 (1923); Sweezy v. New Hampshire, 354 U.S. 234 (1957); Barenblatt v. United States, 360 U.S. 109 (1959); Shelton v. Tucker, 364 U.S. 479 (1960); Griswold v. Connecticut, 381 U.S. 479 (1965).
21. Griswold v. Connecticut, 381 U.S. 479, 482 (1965).
22. Lamont v. Postmaster General, 381 U.S. 301 (1956).
23. Stanley v. Georgia, 394 U.S. 557 (1969).

which human senses and faculties play in making meaningful the scientific process must also come within the ambit of First Amendment coverage. Moreover, when the process is placed in the context of the home—viz., when investigators reflect, tinker, and create (as many of the greats have done) within the confines of their private dwelling places—scientific investigation as a fundamental attribute of the human personality, and therefore as a precious civil liberty, truly comes alive. Were Galileo active today, could public officials enact legislation stopping him from scanning the heavens with a telescope mounted in his private study-observatory? Were Mendel active today, could public officials enact legislation stopping him from testing genetic probabilities through plant cross-breeding in the privacy of his own garden? One does not even have to accept the oft-cited *Griswold* decision—holding that the state lacked authority to ban the use of contraceptives by married couples on the novel ground that "to search the sacred precincts of [their] bedrooms for [such] telltale signs . . . is repulsive to the notions of privacy"[24]—to answer these questions in the negative. And certainly one does not have to accept the Court's recent decisions establishing a nexus between abortion rights and privacy rights[25] to answer these questions in the negative. Galileo and Mendel were searching for nature's laws, not inhibiting through the use of artifacts and medical procedures the procreative process. Moreover, nothing they did or wrote in the name of their investigations involved the manipulation of illicit materials, e.g., obscenity and marijuana. We are dealing with the scientific process as First Amendment "expression" *plus* privacy as an individual liberty formed by the penumbras of specific constitutional guarantees, and it is difficult to perceive a valid governmental interest which would sustain such attempts to "contract the spectrum of available knowledge."

The Mendel analogy introduces a vital element to our understanding of scientific investigation as constitutional value, an element which best characterizes the shift in research strategy from yesterday's emphasis on reflection to today's emphasis on more rigorous exercises. That element is experimentation. When scientists stop talking and start experimenting, some have opined, they move from the world of speech to the world of conduct,[26] and while legislators cannot intrude upon freedom of speech (absent compelling circumstances), they possess wide discretion in regulating action. Clearly, a central issue is whether public officials

24. Griswold v. Connecticut, 381 U.S. 479, 485–86 (1965).

25. Roe v. Wade, 410 U.S. 113 (1973).

26. Hans Jonas, "Freedom of Scientific Inquiry and the Public Interest," *Hastings Center Report* 6 (August 1976): 15–17; Harold P. Green, "The Boundaries of Scientific Freedom," *Harvard Newsletter on Science, Technology, and Human Values,* June 1977, 17-21.

possess the same latitude in enacting ordinances to control experimentation as they have in authorizing codes to govern, for instance, sanitation and zoning.

A review of relevant Supreme Court rulings shows, for one thing, that nice, neat lines separating speech from conduct are largely illusory. A Jehovah's Witness parade is action, but it is also expression; picketing is economic coercion, but it also informs the public about labor's grievances.[27] We are talking at this point of quasi speech, or what court-watchers presently term "speech plus."[28] Indeed, the age-old word-deed dichotomy has become so blurred that there now exists in our constitutional law the notion of "symbolic speech." That is, when students wear black armbands in schoolhouses and antiwar activists burn draft cards in the streets, their protests must be judged not merely as allegedly antisocial behavior that the state has a right to punish but as conduct representing particular points of view. In this context, the rules as applied can be sustained only when "a sufficiently important governmental interest in regulating the nonspeech element can justify incidental limitations on First Amendment freedoms."[29] The test was satisfied with regard to those who destroyed property in which the state had an interest, but it failed to sustain penalties against the pupils, whose demonstrations did not interfere with the rights of others. Of course, a scientist might employ experimentation solely as a means to dramatize some heartfelt opinion, but that simply provides further illustration for the larger point. In cases actually decided by the Court, we learn that while laws cannot ban parades per se, they can ban those parades held near a courthouse with the intent of undermining the fair administration of justice.[30] And while laws cannot ban picketing per se, they can ban those pickets constituting restraints of trade under valid state antitrust statutes.[31] In other words, the "plus" ingredient triggers public welfare considerations that government can address through narrowly tailored rules sensitive to free speech interests.[32]

27. Cox v. New Hampshire, 312 U.S. 569 (1941); Thornhill v. Alabama, 310 U.S. 88 (1940).

28. Note the justices' characterization of speech in the contexts of Cox v. Louisiana, 379 U.S. 559 (1965) and Buckley v. Valeo, 424 U.S. 1 (1976).

29. United States v. O'Brien, 391 U.S. 367, 376 (1968); cf. Tinker v. Des Moines School District, 393 U.S. 503 (1969).

30. Cf. Shuttlesworth v. Birmingham, 394 U.S. 147 (1969) and Cox v. Louisiana, 379 U.S. 559 (1965).

31. Thornhill v. Alabama, 310 U.S. 88 (1940) and Giboney v. Empire Storage, 336 U.S. 490 (1949).

32. A conceptually different situation arises when the state regulates conduct and then imposes sanctions against those who, because they disseminated speech that lacks constitutional standing, have violated the law. For examples, see Schenck v. United States, 249

What then is experimentation? Speech? Action? Quasi speech? To resolve these questions is to resolve the ultimate question whether scientific investigation as a phenomenon of contemporary intellectual life merits First Amendment status. It would seem to follow from everything said thus far that freedom of *expression* embraces not only words but also ideas, attitudes, values, and emotions;[33] that experimentation is "expressive activity," in other words, conduct central to speech; and that such operations are "essential to the exposition of scientific ideas" because they are "the only way researchers can test and substantiate their theories."[34] Does this mean that nuclear physicists have a constitutional right to detonate atomic weapons in search of unknown chemical elements? The standard rejoinder commences with the proposition that policymakers can distinguish between basic research (inquiry) and applied research (technology). It follows that they can also distinguish between basic research that is "inherently hazardous" and basic research that poses hazards only under unusual circumstances. Governmental bodies, the argument concludes, can invoke their police powers to regulate in any rational manner they choose both technology in general and research strategies of inherently dangerous proportions in particular.[35]

To appreciate scholarly enterprise as "expressive activity," when understood in the light of the Court's heightened animosity toward laws which uncritically inhibit quasi speech, is to conclude that scientific inquiry as a generalized phenomenon falls within the capacious scope of First Amendment values. But if one cannot always distinguish, *in the constitutional law context,* between speech and action, how can one distinguish in that context between "pure experimentation" and "applied experimentation," between "pure, benign" operations and "pure, malignant" operations?[36] The literature provides no answers to these questions; much less does it tell us upon whom the burden lies to demonstrate categorical status. Yet these are the salient issues in addressing the constitutional posture of recombinant DNA research. What must be

U.S. 47 (1919) and Arnett v. Kennedy, 416 U.S. 134 (1974). But this doctrine also demonstrates the questionable perception of "speech" and "action" as watertight categories.

33. Melville B. Nimmer, "The Meaning of Symbolic Speech under the First Amendment," *U.C.L.A. Law Review* 21 (1973): 33–34.

34. Delgado and Millen, "God, Galileo, and Government," 378, 380.

35. Ibid., 380.

36. I noted in Chapter 1 the utility of the pure research–applied research dichotomy. But it is one thing to say that the National Science Foundation is involved primarily in the former and quite another thing to say that every study it funds is entitled to an equal measure of First Amendment protection. *Institutional* focus is not necessarily the same thing as *investigative* focus.

done is to examine provocative experimentation as legal scholars have traditionally examined provocative speech, ascertaining *empirically* the nature and likely consequences of *particular cloning investigations.* This assessment would be twofold: (*a*) Is the research essentially either a contribution to the marketplace of ideas or a contribution to some other concern? and (*b*) If the former, is the research of such species that a rational decisionmaker could consider it hazardous?

Gene Splicing and the First Amendment

Biologists have known about deoxyribonucleic acid since 1869, but not until 1944 were they able to mount evidence that it, rather than protein, constituted the fundamental genetic substance. Still, DNA's structure resisted explication until 1953, when Watson and Crick pooled their expertise to show that this large and complicated molecule was, in fact, organized in the aesthetically pleasing, yet eminently coherent form of a double helix. To demonstrate that adenine purines always pair with thymine pyrimidines and that guanine purines always pair with cytosin pyrimidines was to demonstrate in biochemical terms that like begets like through a process of complementary replication, the central principle of heredity for all species. But their discoveries also launched the modern era of genetic research, because they gave investigators a comprehensible unit of material to study, the *base pair.* By this is meant that each gene comprises a string of A-T, T-A, G-C, C-G pairs. A bacterium possesses several millions of base pairs. Theoretically, the human genome—accounting for the makeup of all forty-six chromosomes—is now susceptible of complete mapping and sequencing. We have already learned that the cypher of genetic coding is carried in the nucleotide continuities, that linear combinations of A, T, C, and G trigger specific RNA messenger responses, which, in turn, instruct the twenty amino acids to assume various orderings, and that these aggregations produce protein compounds, the building blocks of protoplasm.

Watson and Crick worked together at a time when the National Science Foundation was barely off the ground, at a time when the United States was setting itself the task of becoming the world's foremost pure research center. Twenty years later, Berg, Cohen, and Boyer each contributed a piece of a puzzle, which, when solved, constituted an incredibly important leap forward in our understanding of the genetic process. Put simply, they showed that base pairs could be recombined, that they could be transplanted into either bacterial or viral base pair systems, where they duplicated themselves following cell division as though they were part and parcel of the host vector. By that time, the American scientific community had indeed achieved a standing second to none, as Nobel

Table 2.1. Nobel Prizes awarded to Americans, 1925–1973

Years	Number of American recipients	Percentage of all recipients by field			
		Physics	Chemistry	Medicine and physiology	All scientific fields
1925–50	26	26%	20%	36%	28%
1951–73	64	53%	35%	54%	49%

NOTE: Again, as with Table 1.1, each time period spans 23 years, and all percentages are rounded down.

Prize data bear witness (Table 2.1). Not surprisingly, the United States was poised to become the cloning capital of the planet.

For some perhaps, the Watson-Crick explorations are a classic example of inquiry, whereas "manufacturing" genetic recombinations via test tube is a classic example of technology. For others, both characterizing the double helix and creating clones from contrived mutants constitute inquiry rather than technology, basic rather than applied endeavors, because they are research performed in the pursuit of knowledge rather than research having immediate practical application.[37] For still others, gene splicing is one of those exercises in contemporary scientific investigation that reasonable people might find, or do find, inherently deleterious.[38] In fact, "recombinant DNA is not a single entity . . . but . . . a group of techniques that can be used for a wide variety of experiments," some safe, some unsafe, some largely a quest for understanding, some largely methods to produce merchandise.[39] Students of law and politics look for nuances in free expression interpretation because they have learned that "speech" is no homogeneous concept. When they assess cloning as being all of a piece, it is very likely because they have yet to study such procedures through First Amendment lenses. I agree that modes of scientific investigation ought not to be treated as quasi speech through some process of self-assertion. A burden of proof lies upon those who claim that particular experimental forms and purposes comprise "expressive activity" and, hence, require some degree of communal solicitude. But the line between gene splicing as a search for truth and

37. Frank Becker, "Law vs. Science: Legal Control of Genetic Research," *Kentucky Law Journal* 65 (1977): 881; John A. Robertson, "The Scientist's Right to Research: A Constitutional Analysis," *Southern California Law Review* 51 (1978): 1203–79.

38. Green, "Boundaries of Scientific Freedom," 19; Delgado and Millen, "God, Galileo, and Government," 380, n. 195.

39. Stanley N. Cohen, "Recombinant DNA: Fact and Fiction," *Science* 195 (February 18, 1977): 654.

gene splicing that lacks pursuit-of-knowledge dimensions can be drawn only by examining specific cases. The research discussed below was selected for illustration only because its objectives are relatively less arcane to the lay reader.

The *Salmonella* are a genus of microorganism whose flagellar control system varies with whether gene H1 or gene H2 is operational. A team of biologists at the University of California—San Diego hypothesized that some sort of shift in the DNA adjacent to the H2 gene accounted for this "switch" phenomenon and contrived a series of cloning maneuvers to establish that fact.[40] First, they employed an appropriate restriction enzyme to cut through the specimen's double helix and isolate relevant genetic sections. Then, they joined those "passenger" substances to similarly cleaved plasmid DNA material that had been taken from the common intestinal bacterium *Escherichia coli*. The vehicle plasmid was subsequently reintroduced into the host, which, following duplication, yielded clones containing the alien genes. These progeny were then run through a series of tests; other recombinations involving the *Salmonella* DNA were also effected. Eventually the researchers located an "inversion bubble" proximate to the H2 gene, sequences of which correlated with the phase variations manifested by that particular gene. The precise nature of these molecular processes remains a mystery; however, the presence and impact of this inversion not only demonstrates that the "switch" effect in *Salmonella* stems from changes in DNA constitution but also promises to provide a fuller understanding of motor control mechanisms in more advanced life forms.[41]

That this research is a bona fide attempt to advance knowledge seems evident. That it deserves to be called "expressive activity" seems equally clear. Yet some *Salmonella* types induce serious illnesses, such as gastroenteritis and typhoid fever. When carried along as part and parcel of the *E. coli* bacterium, an inhabiter of the human gut, the *Salmonella* DNA employed in this research could be imagined to cause medical problems. This danger the state may rationally address through the imposition of appropriate "times, places, and manner" legislation. If communities have the power to punish people who employ the medium of raucous sound equipment to inveigh on the streets,[42] if they can sanction those who demonstrate intentionally on jailhouse property,[43] and who sell sexually explicit (though not obscene) books outside geographical bound-

40. Janine Zieg et al., "Recombinational Switch for Gene Expression," *Science* 196 (April 8, 1977): 170–72.
41. John Abelson, "Recombinant DNA: Examples of Present-Day Research," *Science* 196 (April 8, 1977): 160.
42. Kovacs v. Cooper, 336 U.S. 77 (1949).
43. Adderley v. Florida, 385 U.S. 39 (1966).

aries prescribed by law to quarantine such enterprises,[44] then they can undoubtedly restrict even the most benign recombinant DNA experimentation to certain areas of the landscape and to laboratory facilities meeting minimal standards of cleanliness and safety. To these ends, Cambridge, Massachusetts, insists that certain gene-splicing work be done on premises free from "rodent and insect infestation"[45] and that a local biohazards committee inspect such facilities periodically to check on enforcement.[46] Though this research is conducted at Harvard and MIT rather than in commercial establishments, nondiscriminatory and reasonable "times, places, and manner" rules would appear, on balance, to take precedence over academic freedom in the form of "expressive activity."

Compare the San Diego cloning experiments with Chakrabarty's gene-splicing attempts to create life forms capable of gobbling up oil spillage, described in Chapter 1. Nowhere do the justices, in holding that his artifacts were patentable, discuss recombinant DNA research as free expression, and it is this void that must now be addressed. Recall that, under prevailing interpretations of federal law, newfound abstract ideas and natural phenomena cannot yield privileges of exclusivity. But the question now before us is whether Congress *could,* consistent with the First Amendment, allocate patent rights to Benjamin Franklin for his discovery of lightning as electrical property or to James D. Watson for his conception of genetic chromosomes as double helix. I submit that the legislative branch lacks authority to carve out portions of the idea marketplace and delegate them to the exclusive control of any individual. And the idea marketplace includes formulations of reality, the component parts of the physical and biological order, and the relationships among our countless species. Freedom of expression includes not only the right to talk but the right to talk about any idea or entity, not the right freely to quote those holding copyrights on their forms of words, but the right to debate any question which those writings articulate. Freedom of expression also includes the right to discover what others have discovered, not the right to use their manufactures should those be shielded under the law,[47] but the right to express in one's own way what the world is and the manner through which life forms subsist in it.

Judged against this standard, Congress could no more grant patent

44. Young v. American Mini Theatres, 427 U.S. 50 (1976).

45. Nicholas Wade, "DNA: Laws, Patents, and a Proselyte," *Science* 195 (February 25, 1977): 762.

46. Nicholas Wade, "Gene-Splicing: Cambridge Citizens OK Research But Want More Safety," *Science* 195 (January 21, 1977): 268.

47. As I pointed out in Chapter 1, scientists possess a living constitutional prerogative to employ patentable materials. I do not believe the policy, enlightened though it is, makes

rights to those who isolated the *Salmonella* "inversion bubble" than a state legislature, acting under the Tenth Amendment, could ban such research. From a constitutional law perspective, the discovery of the "bubble" is akin to the discovery of the human heart, and the recombinant DNA process making that discovery meaningful is akin to the modus operandi conceived by a Galileo or a Leonardo to meet objective criteria of proof. If, under the Constitution, public officialdom could not have halted Galileo's or Leonardo's experiments, then it is hard to understand how those experiments, those "ways of knowing," could have been patented. It is true, of course, that had Janssen invented the compound microscope in the United States, he would have been a successful candidate for patent coverage. And so we should not be surprised to learn that when Herbert Boyer and Stanley Cohen applied for such shelter to protect their pioneer gene-splicing methodology, the petition was approved.[48] From this, it hardly follows that Congress could have granted Leeuwenhoek patent rights on his use of the microscope to locate and classify unicellular creatures; nor, to repeat, does it follow that either cloning as a procedure or clones as an end product can be licensed when they are employed in the specific context of idea marketplace exploration. What does follow is that if "science" in the Constitution be seen as fundamental personal activity, then the debate described in Chapter 1 respecting the patent grant as instrument of public welfare now becomes transformed into a debate over whether the patent grant furthers free expression. As we know, Jefferson certainly had his doubts. We need but say at this time that it is for Congress to judge, provided only that the First Amendment process itself is not converted into private repository.

The Chakrabarty enterprises are of a totally different genre. It is not enough, however, to state glibly that his labors were subsidized by a profit-making organization for the purpose of reaping considerable financial rewards and that, therefore, Congress could regulate them or provide patent protection for them as with any other commercial venture. John Bardeen's investigations so significant to the invention of the

---

sound constitutional law. Otherwise, those engaged in free speech (with no commercial consequences) could have used the telephone at their pleasure during the period when Bell's monopoly was enforceable. When Congress provides for patent protection, it does so in a "research neutral" or "speech neutral" context; and neither researchers nor speakers can claim inherent freedom to use someone else's property if that person's property it permissibly can be.

48. In August 1984, the U.S. government awarded Stanford University and the University of California patents covering all hybrid plasmids used to transfer alien genes into prokaryotes, that is, the fruits of the Boyer-Cohen recombinant procedure. It seems clear that if Congress can provide patent protection for particular methodologies, it can accord like privileges to the substances created therefrom.

transistor were accomplished while he was working at the Bell Laboratories, and though he received patent privileges for his artifact per se, it has never been suggested that he could have patented, or that the state could have suppressed, his research into the properties of semiconductors, theoretical notions and experiments upon which the entire transistor industry has been built. In short, it does not necessarily follow that what is pure research at Princeton suddenly becomes applied research at IBM "simply because potential customers for the . . . results existed in the immediate environment."[49] Rather, the locus of analysis must center on precisely what Chakrabarty was trying to do and how he went about doing it. In his earlier endeavors, he advanced our knowledge as to the substance and behavior of bacterial plasmids, research which seems to come well within First Amendment contours and for which he did not apply, and could not have received, patent rights. He then negotiated recombinant DNA processes to create a hybrid bacterium, clones of which would be unleashed solely to clean up the environment at a profit. These experiments revealed nothing new about the nature of our planet or its life forms, nor were they even intended so to do. They may well be consistent with the public welfare, but Congress possesses ample authority either to control or to nurture commercial enterprises by weighing their cost-benefit implications.

To reiterate the central argument: laws distinguishing between cloning as essentially a "way of knowing" and cloning as essentially a means for intentionally producing that which allegedly possesses specific social utility—statutes treating the former with light touches normally reserved for "expressive activity" while treating the latter as substantive evils subject even to state proscription where necessary and proper—are constitutionally permissible. Note that the classical inquiry-technology distinction is still meaningful in this context, but that now we perceive these modes, not as watertight categories, but as functional concepts capable of assuming particular blends depending upon the totality of facts.[50] The jurisprudential task we set is a challenging one. However, it is no more demanding than, and in fact is roughly analogous to, the traditional assessments of fundamental fairness which underlie due process analysis.[51] There, the justices weigh the full range of relevant circumstances in determining whether government officials employed procedures in-

49. Harvey Brooks, *The Government of Science* (Cambridge: MIT Press, 1968), 284.

50. Cf. Loren R. Graham, "Concerns about Science and Attempts to Regulate Inquiry," *Daedalus,* Spring 1978, 12–13.

51. For the Supreme Court's treatment of "totality of facts" as a test for due process fundamental fairness, see Powell v. Alabama, 287 U.S. 45 (1932); Palko v. Connecticut, 302 U.S. 319 (1937); Rochin v. California, 342 U.S. 165 (1952); and McKeiver v. Pennsylvania, 403 U.S. 528 (1971).

consistent with standards of right conduct to deprive a person of life, liberty, or property. It is also roughly analogous to Professor Emerson's "preponderance" test, which he advanced for the purpose of distinguishing between quasi speech falling within First Amendment coverage and quasi speech falling outside constitutional bounds. "The guiding principle," he asserts, "must be to determine which element is preponderant in the conduct under consideration. Is expression the major element and the action only secondary? Or is the action the essence and the expression incidental?" He cites draft card destruction for the purpose of conveying outrage over American military involvement in Vietnam as an instance of the former, while he cites political assassination as an instance of the latter.[52] I have cited as an example of quasi speech the "inversion bubble" cloning experiments and as an example of "nonspeech" investigation the Chakrabarty cloning experiments.[53] Constitutional protection for new forms of scientific exploration and insight deserves no less deference than that which we ought to accord new forms of political protest.

"Buying Up" First Amendment Rights: Gene Splicing as Subsidized Research

The ever-burgeoning public sector has provided recent generations with a variety of services and support structures undreamed of by the Founding

52. Thomas I. Emerson, *The System of Freedom of Expression* (New York: Vintage, 1970), 80–86. For Emerson's comments on the proper scope of governmental power to constrain scientific experimentation, see his 1977 testimony in *Science Policy Implications of DNA Recombinant Molecule Research,* Hearings before the Subcommittee on Science, Research, and Technology of the Committee on Science and Technology, U.S. House of Representatives, 95th Cong., Report No. 24 (Washington, D.C.: Government Printing Office, 1977), 874–90, 905–15.

53. I see little conceptual improvement in labeling applied research as unprotected quasi speech, but I believe Emerson would agree. I also think it analytically unsound to consider *everything* Chakrabarty did as quasi speech and then make an all-or-nothing judgment about whether the First Amendment applies or fails to apply. That process runs the risk of libertarian interest swallowing up social interest and vice versa. Sometimes, however, "pure" and "applied" elements will be inextricably bound together, and only a conventional due process "totality" assessment will suffice. These experiments, no matter on which side of the line they sit, I cheerfully agree to call quasi speech.

I part company with Emerson on the difficult "burden of proof" question, though, probably because political action is easier to judge by conventional standards of evidence than is scientific action. Surely the "preponderance" approach is sufficient for ascertaining whether conduct is, in fact, *scientific* conduct, viz., whether alleged experimentation is really what it purports to be. But once it is found that science is involved, an investigator has the burden, I think, of establishing a prima facie case of constitutional conduct (cf. the comments in text following n. 39). As a matter of living constitutional expectations, a *grant* from NSF or NIH for cloning research would probably be sufficient to meet this burden, though the presumption would be rebuttable.

Fathers. As Figure 1.2 graphically shows, scientists have benefitted enormously from that development. Without the state to aid and abet their efforts, the high cost of contemporary experimentation would put them at the mercy of private foundations, universities, and corporations, which, in 1978, funded but 31 percent of the basic research conducted in this country.[54]

Government sponsorship, though, invariably means government strings. And sometimes those strings include even deprivations of constitutional liberty, as the Supreme Court has frequently decided. At one time public officials attempted to keep "subversive speech" from dissemination through the mails.[55] They also tried to abort welfare payments without providing either a hearing to the deprived beneficiary or an exemption for those whose freedom to practice their religion would be compromised if they accepted work on their holy days.[56] They even tried to force public school teachers to sign a certificate upon pain of dismissal that they did not hold membership in communist organizations.[57] Yesterday, when public and private spheres were nicely demarcated, judges could argue that the state had power to fire policemen who delivered political speeches because the Constitution does not guarantee a person's right to serve on the police force.[58] Today, when these zones shade together, judges are far more apt to argue that government can create or abolish the postal service, create or abolish welfare, and create or abolish its own schools at will, but as long as these facilities and programs abide, those to whom they cater and those who work to keep them operational must be accorded whatever constitutional rights they possess in their more traditional relationships with the state. And so if scientific investigation is an aspect of free expression, then to what extent can government supervise such investigation on the ground that it provides the research dollars? An analysis of relevant statutes, guidelines, cases, and doctrinal principles yields the conclusion that there is a middle ground between the above polar positions: viz., government lacks constitutional authority as funding agent to place whatever conditions it desires on research, but the state does possess considerable leeway in deciding how those monies should be expended, a range of discretion

54. National Science Board, *Science Indicators—1978,* NSB-79-1 (Washington, D.C.: Government Printing Office, 1979), 179.

55. Milwaukee Publishing Co. v. Burleson, 255 U.S. 407 (1921). But see also Justice Holmes's dissent in that case. The modern view is expressed in Lamont v. Postmaster General, 381 U.S. 301 (1956) and Blount v. Rizzi, 400 U.S. 410 (1971).

56. Goldberg v. Kelly, 397 U.S. 254 (1970), but cf. Mathews v. Eldridge, 424 U.S. 319 (1976); Sherbert v. Verner, 374 U.S. 398 (1963).

57. Keyishian v. Board of Regents, 385 U.S. 589 (1967).

58. McAuliffe v. Mayor of New Bedford, 155 Mass. 216 (1892).

which in the context of criminal law enforcement would not be deemed permissible.

There are a number of circumstances in which federal authorities have constrained scientific activity under taxpayer auspices without causing arguable constitutional difficulties. One such instance involves national defense. During the 1930s, the discovery of radar was kept under wraps for security reasons, while, in 1946, as noted earlier, Congress enacted legislation withholding nuclear research from the public domain. Surely, these sorts of restrictions are entirely appropriate to federally sponsored work, no matter who the principal investigator and no matter where the project is housed. Moreover, the state can lay down rigorous standards of access to sensitive materials in its possession. The issue here centers on, not a constitutional power to classify data, which is beyond cavil, but rather, a constitutional prerogative to disqualify personnel from ongoing experimental endeavors on grounds of disloyalty, moral turpitude, and the like. Perhaps the Atomic Energy Commission acted improvidently when it branded J. Robert Oppenheimer a security risk, but that he could be forced to surrender research entree where the state demonstrated reasonable grounds in a context of fair procedure seems unassailable. And certainly scientists investigating biomedical questions at the National Institutes of Health must satisfy whatever bona fide research guidelines Congress might care to impose. That is, government employees cannot claim freedom of inquiry when their superiors order them to look elsewhere for the right answers.

But matters become less clear when we approach the subject of direct concern here, the investigator who works usually in the academic environment and who has applied for a federal grant to study some question in basic research. Upon this much, all agree: government agencies are under no constitutional duty to fund either scientific activity in general or the natural sciences, physical sciences, and social sciences as discrete categories in particular. And these agencies are free to pick and choose among disciplines subsidized: they can select history but reject sociology, encourage biology but ignore geology, assist botany but refuse to assist zoology. This is, of course, a form of discrimination, wherein the state designates for subsidy some aspects of knowledge, some aspects of the idea marketplace, but declines to promote other aspects. Yet, just as the public schools do not have to teach everything in order to teach something, so public-funding instrumentalities need not give money to everyone in order to give money to someone. Even though all these disciplines and fields are in the nature of speech, the distinctions made between and among them are consistent with the equal protection tenet because they are rational; and the constitutional test is one of rationality

—rather than one which forces the state to demonstrate compelling interest—because there is only so much money to go around and because the categories drawn do not disfavor some notions of truth as against competing conceptions of truth. For when the Framers adopted the free expression principle, their central goal was to preserve and foster a content-neutral marketplace of ideas, one in which national government abjures from making official findings of truth and employing whatever political levers are at its disposal to drive competing values from that marketplace.[59] And that principle applies to quasi speech as well as to pure speech, to "times, places, and manner" rules as well as to censorship laws.[60]

Let us suppose, however, that Washington has decided to fund biomedical investigations and that Professor Smith's grant proposal receives support. What obstacles in the form of strings will he encounter, and to what extent are these conditions subject to challenge? Again, the proposition is incontestable that permission to perform biomedical studies is not permission to perform exercises above and beyond the scope of biomedical analysis, and it most certainly does not include any justification for falsifying one's results or manipulating in some other fashion the experimental process so as deliberately to taint the search for truth. Unfortunately, the current academic environment with its emphasis on tenure through publication—or perhaps one should say tenure through *number* of publications—is sometimes not conducive to the painstaking deployment of scientific expression. In one well-publicized but not atypical case, a Harvard Medical School investigator working on an NIH-sponsored project confessed to fabricating evidence and was stripped of his hospital appointment. However, the fact that Harvard allowed him to remain a part of the research team at the taxpayers' expense for six months thereafter, and reached that determination without even informing NIH of the scientist's admitted perpetration of fraud until some time later, enforces the belief that in such instances the danger of academic cover-up is much greater than the danger to academic freedom.[61]

59. Cf. Delgado and Millen, "God, Galileo, and Government," 389 n. 237. The Court's classic statement reads as follows: "[The state cannot] restrict expression because of its message, its ideas, its subject matter, or its content." Chicago Police Department v. Mosely, 408 U.S. 92, 95 (1972).

60. Carey v. Brown, 447 U.S. 445 (1980); Erznoznik v. Jacksonville, 422 U.S. 205 (1975).

61. William J. Broad, "Report Absolves Harvard in Case of Fakery," *Science* 215 (February 12, 1982): 874–76. The report referred to here resulted from an investigation by a committee five-eighths of whose members were drawn from the Harvard faculty. For Yale's contribution to the research scandal literature, see Morton Hunt, "A Fraud That Shook the World of Science," *New York Times Magazine,* November 1, 1981, 42–75.

Our Professor Smith, of course, will be asked to prepare the usual forms and progress reports one associates with Washington oversight. But the bane of researchers supported by federal grant money is OMB Circular A-21, which, in its revised 1979 edition, mandated principal investigators to fill out "effort reports."[62] If these information sheets had required only explanations of how much time one spent "on project," they might have been considered merely a case of good intentions having gotten somewhat out of hand. But OMB wanted to know how faculty members apportioned 100 percent of their on-the-job labors. To be sure, the obligation was easily satisfied; scientists needed only specify approximate breakdowns among teaching, research, and administrative responsibilities. Some investigators, however, considered the reports an invasion of privacy, bearing little relation to sensible checks for accountability.[63] If the constitutional right of privacy means anything, it is conceivable that there are some things government agencies cannot expect us to answer even on so-called effort reports. But there is a rational relationship between knowing how much time we put in on various aspects of our work assignments and how much time Washington expects us to put in on research it funds. One is forced to confine one's criticism of this meddlesome, rather infantile exercise in red tape to the observation that it was just that. In 1982, OMB again recast A-21, striving at that time to provide only for feedback on "work funded by the federal government," though criticism has certainly not abated regarding the "opaque details" required.[64]

Even Washington's authority to classify sensitive research explorations funded at public expense, which I earlier said was "beyond cavil," can be strained to the point of rupture in the project-grant context. In April of 1982, President Reagan issued an executive order giving federal officials the power to classify for an indefinite period basic scientific information "clearly related to the national security," even where the scholarship had not been classified at the time it was approved and subsidized.[65] So a Sword of Damocles hangs over faculty members who, armed with NSF or NIH grants, pursue path-breaking cloning investigations in their academic settings: if federal officials perceive, after the fact, an unmistakable nexus between a study and the nation's defense,

62. "Cost Principles for Educational Institutions," *Fed. Reg.* 44 (1979): 12368.

63. Colin Norman, "Yale Professor Refuses to File Effort Reports," *Science* 215 (January 15, 1982): 274; Norman, "Faculty v. OMB: One More Time," *Science* 215 (February 5, 1982): 642.

64. Colin Norman, "A-21 Rules Take Effect," *Science* 217 (August 27, 1982): 810.

65. Robert A. Rosenbaum et al., "Academic Freedom and the Classified Information System," *Science* 219 (January 21, 1983): 258. For an update on implementation, see John Walsh, "DOD Springs Surprise on Secrecy Rules," *Science* 224 (June 8, 1984): 1081.

publication of research results may die aborning unless government censorship agencies conclude otherwise. Though no such instance has yet occurred, one must question the constitutionality of retroactive classification for these pure science exercises, absent either exigent circumstances (e.g., the specification of particular harm) or procedural due process.

To this point, we have not considered regulations and guidelines structuring the *content* of legitimate scientific inquiry. Putting to one side for the moment NIH constraints on cloning itself, the matter of greatest relevance for our purposes is research entailing human subjects. Without doubt, the state has ample authority to protect persons as data from persons as investigators. In other words, a scientist's freedom cannot weigh more heavily than the individual's dignity as experimental participant from the standpoint of constitutional right, and government can strike, within the bounds of reason, a balance between these two interests. Thus Congress has by statute provided that NIH-funded experiments involving humans cannot go forward unless subjects eighteen years and older first tender their "written informed consent," while those of younger age must obtain a similar waiver from either parents or legal guardian.[66] Actually, this sort of oversight very nearly approximates a "times, places, and *manner*" check, for it leaves researchers free to explore what they will, and to indulge the expressive activity of their choice, provided the subject takes an affirmative, meaningful step forward. A constitutional problem would arise only should the state attempt to proscribe types of experiments concurred in by both parties, a form of paternalism I now address.

According to Department of Health and Human Services (HHS) guidelines, pregnant women fall into a special class, i.e., research directed toward their particular condition and employing them as subjects shall be approved for funding only when (*a*) the purpose is to address the participant's health requirements, in which case the fetus must be "placed at risk only to the minimum extent necessary," or (*b*) the purpose is something else, in which circumstance "the risk to the fetus [must be] minimal."[67] So no matter whether pregnant subject and principal investigator think otherwise, HHS as research benefactor introduces a "larger interest" into their transaction, concern for the *potential* human being, the fetus. I emphasize "potential" because the Supreme Court has specifically found that a fetus is not a person in the sense that its life, liberty, or property remain inviolate absent due process.[68] And yet,

---

66. PL 94-206, sect. 411; 90 *Stat.* 25 (1976). See also U.S. Department of Health and Human Services, *NIH Extramural Programs,* NIH Pub. No. 80-33, July 1980, 9.

67. 45 CFR 140, sect. 46.207.

68. Roe v. Wade, 410 U.S. 113 (1973).

caveat *a* also has the ring of a "manner" regulation, for any and all agreed-upon research relevant to the mother's health needs is acceptable so long as there is but a "minimum infringement," given research purposes and objectives, to the fetus.[69] This guideline, then, owes an intellectual debt to those cases where the Supreme Court allows the state to regulate the "plus" of quasi speech if the inhibitions upon speech itself are as narrowly drawn as conditions warrant.[70] That is, a pregnant woman herself occupies a sort of quasi status under the law, and the value of the fetus—not to her but in its own right as potential human—can be analogized to the value of First Amendment discussion in our cultural universe. But we also learn that the analogy breaks down at the level of total proscription, for while speech can sometimes be halted in the name of larger social evils posed by the "plus," a congressional enactment specifically precludes NIH-supported scientists from inducing an abortion to further *any* scholarly interest related to family planning or population research.[71] Moreover, caveat *b* as well authorizes limits on scientific inquiry per se, for, putting aside maternal health needs, some investigations will surely be off limits because the fetus cannot be accorded maximal protection. If we could agree that potential life is as worthy of constitutional shelter as the most pristine forms of scientific expression, then these two regulatory red lights, one legislative, the other administrative, certainly would require no further defense. But that is a proposition which is impossible to square with *Roe v. Wade* if experimentation as First Amendment value means anything. Moreover, just as there is a significant distinction between therapy and research,[72] so the significant distinction between basic and applied investigations does not necessarily evaporate simply because the unit of analysis is the pregnant woman. Specifically, it is one thing to place these human subjects on varying diets for the purpose of studying their metabolic responses as compared to those of nonpregnant females; it is quite another thing to ply the subject with a particular drug for the purpose of seeing whether the artifact lessens pain in delivery, as the inventor-doctor says it will.[73]

69. Exception: where the activity entails termination of the pregnancy. 45 CFR 140, sect. 46.206.

70. Cf. United States v. O'Brien, 391 U.S. 367 (1968), and Martin v. Struthers, 319 U.S. 141 (1943). For a discussion of the relationship between the "minimum infringement–least means" test and "times, places, and manner" rules, see Martin Shapiro, *Freedom of Speech* (Englewood Cliffs: Prentice-Hall, 1966), 140–43.

71. 42 USC ch. 6A, sects. 1004 (b) (2) and 1008; *NIH Extramural Programs,* July 1980, 9.

72. See Bernard Barber et al., *Research on Human Subjects* (New York: Sage, 1973), 59ff.

73. This analogy is suggested by the commentary in Bradford H. Gray, *Human Subjects in Medical Experimentation* (New York: Wiley, 1975), 56–59.

The tension between a researcher's initiatives and the fetus as research subject is drawn into bolder relief by other HHS funding reservations. Assume we are discussing potential life in utero, where all agree the mother's interests are at their highest. Investigations of this kind are banned unless (*a*) the purpose is one of fetal health and the above-stated minimum-infringement criterion is satisfied, or (*b*) the objective is "the development of important biomedical knowledge which cannot be obtained by other means" and risk to the fetus is minimal.[74] Note that the former is somewhat different conceptually from provision *a* for research with pregnant women, because it posits a relationship in which the prospective mother is no factor at all except as consenting agent. To that extent, it is far less controversial in theory. Item *b,* however, is *more* controversial than provision *b* in the pregnant female context. In order to pass muster, basic science inquiries must be "important" and fetal research must be the *only* medium available for such "expressive activity." The notion that a fetus deserves greater protection under the law than a female with child, when the Supreme Court in *Roe* has weighted the scales in diametrically opposite fashion, can only be explained politically. The question is whether it can be justified constitutionally. Below, I discuss Medicaid regulations restraining, with but few exceptions, the abortion process. There, however, the issue is not procreation plus scientific analysis; it is procreation pure and simple. Here, HHS is saying to biomedical scholars: if you want to study fetuses in utero (where, after all, most of them can be found) and if the mother knowingly concurs in both your objectives and your strategies, then the project can go forward only should we agree that the research is significant, that the proposed methodology is the only way to obtain the desired knowledge, and that the fetus will be handled at minimal risk. This guideline is constitutionally suspect, unless one is prepared to argue that HHS can lay down such rules because it exercises total control over its own purse-strings. As was implied above, that proposition is fraught with jurisprudential complexity, and I shall attempt a clearcut refutation after I have subjected the Court's abortion-funding pronouncements to careful scrutiny.

We are now prepared to ask the most important question in this phase of our analysis: can the state choose to support experiments in genetics while refusing to support experiments employing recombinant DNA? This essay has argued that an outright ban on cloning would abridge fundamental freedoms, and some libertarians are of the opinion that government has no greater authority to introduce "content distinctions" into the funding decision than it has to invoke them in framing criminal

74. 45 CFR 140, sect. 46.208. For the far less frequent investigations involving fetuses ex utero, see sect. 46.209.

statutes. With due regard to Warren Court arguments advanced in the *Sherbert-Keyishian* line of cases,[75] the suggested congruency treats a delicate constitutional question with a sledgehammer rather than with a scalpel. For one thing, research proposals fail of approval every day of the week because their content seems misguided or inane. Suppose the National Science Foundation puts up money to support grants in "science and society," and someone requests funding to prove that cloning is a communist plot or a capitalist swindle. Have the researcher's rights been transgressed because NSF officials think what he has to say is prima facie spurious even though he has a right to say it? Hardly. For another thing, there are significant Supreme Court precedents to the contrary. When workers take jobs with the federal civil service, Congress can force them to surrender their constitutional right to engage in partisan political activity.[76] And when indigents receive money from the federal treasury to defray medical expenses, Congress can forbid them to spend any of it on abortions, unless the operation is either necessary to save the life of the mother or to prevent offspring arising from rape and incest.[77] These qualifications severely undermine the constitutional privacy right of poor women to terminate their pregnancies during the first two trimesters, and they are qualifications based solely on the government's dislike for the moral and ethical *content* of a choice these women may well approve. The lesson seems clear. When government wants to ensure an appearance of neutrality among its employees, it can compel them to steer clear of political rough-and-tumble. When government wants to fight hunger, it can dispense welfare checks in the form of food stamps to make sure the largesse is not spent on something else, for instance, on exercising free speech. And when government wants to spend its own money protecting fetuses, it can refuse to give wherewithal to women whose definitions of life and whose theories of procreation are perhaps not consonant with the state's own. Still, there remains the argument that public officials can fund abortions or not fund abortions, but that once they decide to open the door they cannot impose content-oriented criteria.

It is instructive to compare the recent debate over creationism in the public schools with the abortion-funding controversy. Arkansas had enacted a Balanced Treatment statute, taking the position that a state, in managing *its* own schools, might not be able to block the teaching of

75. Sherbert v. Verner, 374 U.S. 398 (1963); Keyishian v. Board of Regents, 385 U.S. 589 (1967).

76. Civil Service Commission v. National Association of Letter Carriers, 413 U.S. 548 (1973). Also on point is Healy v. James, 408 U.S. 169 (1972).

77. Harris v. McRae, 448 U.S. 297 (1980); cf. Maher v. Roe, 432 U.S. 464 (1977).

evolution but could certainly authorize equal time for creationism. This logic the courts rebuffed, citing the establishment of religion principle.[78] But suppose the state's cherished doctrine had lacked theological overtones. Would the Constitution have permitted Arkansas to tell evolutionists: "If you don't want to hear about creationism, then repair to your own schools." I think not. A central theme of academic freedom, a central theme of research inquiry in the classroom or the laboratory is that teachers should be able to impart scientific knowledge as they perceive those truths, assuming of course minimum standards of expertise. That a school board can require a course in biology does not give it the power to require a course on evolutionism-creationism in the presentation of which the scientist-teacher is told to give a predetermined measure of deference to both versions of natural history. Content neutrality is not the same thing as coerced content equality.

How does this argument help us understand what the Court has said in the abortion cases and what it ought to say, if ever asked, in the gene-splicing context? For our purposes, the justices submitted three findings in constitutional law to uphold Congress' abortion-funding guidelines. The first point was simply that the state's impotence to limit unduly the abortion transaction in no way created an affirmative taxpayer duty to finance that transaction. This contention we may dismiss summarily, because here the state had decided to open the door and subsidize the abortions of its choosing. The second point was that Congress' standards left aggrieved indigent women where they stood before the gifts were tendered to others, in no way disparaging the alternative private sources normally available. Just how many private clinics would be prepared to service the needs of these indigents is not clear. At least evolutionists (and creationists) have a reasonable likelihood of fending for themselves if they find public education repugnant. But of equal significance, why must the typical indigent woman who wishes to terminate an unwanted, and perhaps even an unhealthful, pregnancy be thrown onto her own resources when evolutionists can rely on the courts for help? Perhaps it is because the state interest at issue is different—i.e., Arkansas cannot weight creationism equal to Darwinism, but Congress can favor childbirth over abortion to whatever degree is considered rational. And this really is the justices' third point; indeed, it is the crux of the entire matter. I submit that *Roe v. Wade* created a constitutional right to obtain an abortion, not an *absolute* liberty, for freedom of speech itself is not absolute.[79] I further submit that Congress can no more pick and choose

78. Roger Lewin, "Judge's Ruling Hits Hard at Creationism," *Science* 215 (January 22, 1982): 381–84.

79. Though in *Roe*, the Court said specifically that the right to terminate one's pregnancy during the first trimester *is* absolute.

among competing theories of procreational choice which justify in various doses the decision to terminate a pregnancy than it can pick and choose among competing ideologies in the marketplace of ideas. The justices have vested procreational activity with a protective constitutional mantle,[80] and the fact that quasi conduct (as Emerson might term it) is involved has in no way vitiated the strength of the libertarian interest at stake according to these pronouncements. The Court's thesis in *Harris v. McRae* comes down to the implicit notion that expression is a "preferred" right, while aborting the fetus in the name of personal privacy is a "deferred" right. I question this hierarchy of constitutional freedoms. Moreover, even should one accept *arguendo* the validity of that supposition, the evidence now becomes overwhelming that when scientific inquiry as First Amendment freedom conjoins with the woman's decision to place her unborn child at the investigator's disposal, HHS's selective criteria for in utero fetal research are unconstitutional.

Evidently, the justices are groping toward a functional test for constitutional funding strings, a test eschewing either-or distinctions between rights and privileges, between benefits carrying no permissible qualifications and benefits carrying any and all permissible qualifications. Putting aside the question of whether the Court has been consistent in its application, or has subjected to scrutiny respective variables as we might have assayed them, this test may be articulated as follows: where government seeks to promote a valid legislative/administrative goal and where withholding benefits is rationally related to that goal, public officials may attach strings, provided such standards (*a*) do not discriminate in favor of the majority at the expense of "insular" groups; (*b*) do not favor some notions of truth over competing notions; (*c*) do not lump together uncritically as bringing about "substantive evils" various attributes of constitutional right, only some of which require political oversight; (*d*) do not make it inordinately difficult for those deprived of largesse to obtain assistance elsewhere; and (*e*) do not strip recipients of meaningful liberty or property interests without commensurate procedural guarantees.[81]

---

80. Carey v. Population Services, 431 U.S. 678 (1977).

81. A few rulings cannot be squeezed into this formula, largely because, as we have seen, the state sometimes can demonstrate not merely a valid supervisory interest, but an overriding or compelling regulatory concern. One such context is that of national security. Another, as exemplified by Wyman v. James, 400 U.S. 309 (1971), involves the protection of dependent juveniles whose standing to receive public assistance might be jeopardized by irresponsible adults. A third area is that of tax exemption qualifications, where, in Bob Jones University v. United States, no. 81-3 (1983), the justices upheld an IRS holding which stripped a whites only institution of its tax exempt status on the ground that racial discrimination violated "fundamental national policy." For the state to afford private schools financial relief unless those schools base their admissions policies on racial (rather than on, say, religious or other doctrinal) preferences appears to be a clear case of content bias.

Against the backdrop of this doctrine as well as my assessment of how it can be applied fruitfully to new constitutional horizons, I herewith assert that Congress (or its bureaucratic agents) could decline to foster cloning were there reasonable grounds for finding such experimentation hazardous. Moreover, Congress could decline to subsidize cloning directed toward producing goods which it might rationally consider harmful. But to reassert the central point: recombinant DNA research is a generic term encompassing a wide variety of experiments aimed at accomplishing a wide variety of results. What rational purpose would be served and what valid legislative goal would be effectuated where government subsidizes investigations into the nature of genetic material but refuses to subsidize "expressive" probings of the *Salmonella* H2 gene simply because the state has labeled cloning per se, without any empirical evidence whatever, a species of "forbidden content"?[82] To place such loosely structured conditions on funding scientific inquiry is indeed to resurrect the dogma that if the state can keep policemen from managing political campaigns, then it can also keep them from making public their political views.

One last question awaits analysis: what strings can government place on recombinant DNA research selected for financial support? The next chapter delineates the core provisions of the NIH guidelines themselves and discusses their constitutional justification. The following remarks, then, address only the salient theoretical parameters. First, rule-makers possess undoubted authority to insist upon *reasonable* "times, places, and manner" criteria with respect to how expressive activity shall be conducted. Second, public officials cannot exact as a price for funding the restraint of publication. The whole purpose of cloning as basic research is to develop insights which can be disseminated in the idea marketplace. A Berlin Wall blocking exchanges in information gained from expressive activity would stand only if the state could show valid national security interests. There is no evidence that anyone can make a recombinant DNA bomb or unleash a recombinant DNA monster. Third, government goes too far where it subsidizes cloning exercises but then attempts to dictate what experiments should be conducted and what results should be achieved. Granted, funding agents can expect researchers to file comprehensive, even very detailed, strategic designs; oftentimes relevance and competence, not to mention originality, cannot be evaluated until these blueprints are digested. And, of course, "manner" considerations enter here as well, for the use of certain equipment or specimens can pose

82. Of course, the San Diego proposal could fail the quite different content-oriented test described above, namely, that the project be meritorious. Needless to say, no reasonable person would contend that recombinant DNA research is per se unworthy.

special problems. Still, the state lacks authority, even as gift-giver, to stack the content deck. The cloner must retain a range of discretion to employ all pertinent modes of inquiry and to address all pertinent questions of substance within the bounds of the context agreed upon. And within those bounds, it is the gene splicer who must, in the end, define pertinence. Fourth, government cannot reserve for itself unlimited power to strip investigators of their allotted research prerogatives. The Supreme Court has said that public agencies have a duty to provide a hearing on the issue of fault to most students before suspending them from school, and these agencies are also constrained to provide a hearing on the issue of fault to most motorists before suspending their drivers' licenses.[83] I contend that when grants are found subject to termination because of questions of personal responsibility and accountability, recombinant DNA scientists are entitled to some measure of procedural due process. In these various ways, the taxpayer, through elected and appointed delegates, can encourage an extraordinarily richer exploration of science's newest frontier, consistent with the public welfare and yet honoring the fundamental attributes of research liberty. Such an environment will guarantee investigators the freedom to develop inquiries that are not only bona fide basic science but also meaningful expressions of their artistic temperaments.[84]

83. Goss v. Lopez, 419 U.S. 565 (1975); Bell v. Burson, 402 U.S. 535 (1971), as refined by Dixon v. Love, 431 U.S. 105 (1977).

84. In his "The Life Sciences and the Public: Is Science Too Important to Be Left to the Scientists?" *Politics and the Life Sciences* 3 (August 1984): 30–31, Malcolm L. Goggin accuses me of arguing that the right of laboratory experimentation is an absolute. My only rebuttal is the message of this chapter, which is almost precisely the message I published in 1981.

3 The Recombinant DNA Debate as Constitutional Dialogue

Path-breaking science, by its very nature, contains the seeds of culture shock. Galileo's ideas eventually altered the ways in which the Western world would define man's sense of belonging; Freud's theories, in time, changed the manner in which Western civilization would conceive the mainsprings of human behavior. And sometimes the shock is that much greater because the medium is part of the message. Joyce could not tell us about his Dublin and Faulkner could not tell us about his South without invoking new images through unique grammatical and stylistic modes. Watson searched for the structure of DNA his way; Rosalind Franklin searched her way. It is not necessarily a question of the right way, because often there are many ways. In our time, some of those procedures—psychosurgery, manipulation of human fetal materials, computer technology, lasers, nuclear fission, recombinant DNA—challenge the cultural ethos as dramatically as did the substantive notions of those whose wisdom we now consider a precious legacy. So the recombinant DNA debate is a social phenomenon prompted by a unique methodology and potentially unique outcomes. American politics, like all politics, strains to reconcile new lines of inquiry, new formulations of truth, with the established order. Innovative theories, sometimes explaining, sometimes justifying these accommodations, are necessitated, as the contents of the preceding chapter make manifest. But the process of effecting an appropriate balance itself poses severe tests for the political system. That is because conceptions of public welfare, of governmental prerogative, of personal freedom, of group identity, most likely will dictate the manner in which society responds to the challenge of newness, to the challenge of scientific breakthrough; and the clash of attitudes toward these notions as well as the clash of interests in making those attitudes

60

preeminent can cause a drastic rethinking about the nature of relevant constitutional norms.

This chapter speaks to the recombinant DNA debate as a constitutional process. It attempts to show how the political rules of the game have structured that debate and how it, in turn, has influenced those rules. We will see that science is capable of causing not only culture shock but constitutional shock, and we will probe both the dynamics and the consequences of these tremors in political consciousness.

In the twenty-year period following Watson and Crick's discovery of the double helix, molecular biologists learned a great deal about genetic sequencing. They determined that every three links on each of the nucleotide chains (e.g., T-A-T, C-A-T) was code language activating an amino acid, and they also ascertained that strings of these "triplet codons" caused the twenty amino acids found in all life forms to arrange themselves in various structures, each of these formations constituting a protein. But if each cell of a human being contains the same double helix, what is it about genetic expression that causes cell differentiation? Needless to say, one could ask the same question about any other organism, and many scientists in the early 1970s were doing just that. A leading geneticist engaged in these inquiries was Stanford's Paul Berg. He would contrive one of the first recombinant DNA experiments. He would also precipitate the recombinant DNA debate. What was his intent?

A virus, when it invades living tissue, usually causes cellular malfunction by instructing the host creature to express not only its own DNA but viral DNA. An especially well-charted organism is the monkey virus SV40. This virus was a frequent natural contaminant of the early polio vaccines developed by Salk, but there never has been evidence that it caused adverse effects on the countless children treated with such strains. The SV40 contains only seven genes, and if it could be introduced into a bacterium, then the researcher would have a most serviceable guinea pig for studying patterns of genetic function leading to tumor genesis. Berg's experimental design was, first, to cleave the SV40's double helix; second, to cleave the double helix of another virus, an antibacterial agent known as bacteriophage lambda; third, to fasten the former to the latter; and, finally, to place the mutant genetic material into a laboratory strain of the *E. coli* bacterium, which was expected to produce cloned SV40 DNA.[1] Stages 1–3 he proceeded to achieve; hence, the scientific community could boast of mastering a recombinant DNA procedure of signal importance. But the SV40 was known to cause tumors in mice,

1. Professor Berg's Nobel Prize acceptance speech, in which he describes his research, can be found in *Science* 213 (July 17, 1981): 296–303.

though obviously not in monkeys, and *E. coli* (though not the strain he used) was an inhabiter of the human intestinal tract. Suppose Berg's clones somehow escaped into the environment or were picked up by laboratory workers, and infected recipients found themselves cancer victims? Upon the pleas of several fellow investigators, Berg held the concluding step of his research in abeyance. Also staying his hand was Stanley Cohen, who had discovered that *E. coli* plasmids made excellent vectors for genetic recombination, and who, utilizing the restriction enzyme cleavage technique he and Herbert Boyer would later patent, had actually employed some toad DNA to that end.[2] We see here peer review being mobilized not for the purpose of checking the quality and competence of the science but, rather, for the purpose of monitoring the hazards allegedly attendant upon the work. And in this instance, the upshot of peer review had been an old-fashioned case of self-censorship.

The next development in the unfolding politics of cloning took the form of a letter from leading researchers to the president of the National Academy requesting that he appoint an ad hoc committee to study biosafety ramifications.[3] This strategy was hardly unprecedented, as NAS on several occasions had undertaken expert evaluation of controversial scientific activities, some of these entailing public policy consequences at least as provocative as those seemingly at issue here. For example, during the Eisenhower presidency, a National Academy of Sciences committee looked into alleged hazards stemming from atomic testing, concluding, much to the consternation not only of "ban the bomb" activists but also of a good many watchful researchers, that radioactivity resulting from the U.S. defense program was a paltry contribution to the overall level of contamination.[4] The recombinant DNA panel, chaired by Paul Berg himself, made two significant determinations. First, it concluded that the matter was entirely too big for committee dispensation; an international conference of specialists should be convened. Second, until that time— an interim period of about ten months, as it turned out—other scientists were asked to observe the kind of moratorium Berg and Cohen had

2. These comments follow Clifford Grobstein, *A Double Image of the Double Helix* (San Francisco: Freeman, 1979), 16–18. Also of assistance is Frederic Golden, "Shaping Life in the Lab," *Time,* March 9, 1981, 54, 58. An in-depth calendar tracing the twists and turns of recombinant DNA disputation during the 1970s can be found in James D. Watson and John Tooze, *The DNA Story: A Documentary History of Gene Cloning* (San Francisco: Freeman, 1981). For analysis of these events which differs radically from the picture I present, see Sheldon Krimsky, *Genetic Alchemy: The Social History of the Recombinant DNA Controversy* (Cambridge: MIT Press, 1982).

3. For this communication, see *Science* 181 (September 21, 1973): 1114.

4. J. Stefan Dupré and Sanford A. Lakoff, *Science and the Nation* (Englewood Cliffs: Prentice-Hall, 1962), 129.

placed on themselves.[5] Whether this request was honored by all concerned is not clear, but without doubt, anyone who bucked the tide did so at great professional risk and took extraordinary care to keep his investigations secret.

As Grobstein has pointed out, there are apparent parallels between the plight of atomic scientists in the late 1930s and the agonies of gene splicers in 1974. Both research areas offered enormous potential for changing the very nature of human existence.[6] But the analogy is certainly far from a perfect one. As we have seen earlier, Hitler's rise to power coupled with Pearl Harbor made it virtually mandatory for American nuclear physicists to drape a censorship blanket around their research findings and, later, to join in the race to produce an atomic bomb. If national security interests have ever placed special duties upon science, those years will serve as the archetypal standard. Even then, however, the NAS probably overstepped the bounds of propriety—and certainly overstepped its constitutional authority, if public agency it be— when one of its committees tried, in 1942, to stop the publication of an article detailing the role of penicillin as disease fighter.[7] The point is, of course, that not even the war power accords federal officialdom a blank censorship check: scientific discourse regarding penicillin bears no relation to classified data on nuclear testing. And just what comparison should be drawn between Paul Berg's purpose of discovering patterns of genetic expression and J. Robert Oppenheimer's purpose of constructing the "ultimate" weapon? The one was a classic case of expressive activity, a form of protected constitutional conduct, which could easily have been carried out consistent with reasonable "times, places, and manner" caveats and which was, furthermore, to have been effectuated in an academic setting. The other was a classic case of war-making activity, necessary for national defense but hardly protected under the First Amendment. If we are looking for parallels, though, we should not overlook the year of recombinant DNA decision: 1974. There was no Hitler, but there was Watergate. "Let's get it all out in the open," the relevant molecular biological community must have thought. "Let's have maximum participation. We couldn't have an international assembly in 1940 discussing atomic secrets, but we can have one now. When in doubt, caution should prevail; but by all means, let's not let anyone accuse us of a cover-up."

The Asilomar (California) Conference was held in February 1975 and

5. The committee's recommendations were made generally available in *Science* 185 (July 26, 1974): 303.

6. Grobstein, *Double Image*, 21.

7. Letter from Michael M. Sokol published in *Science* 215 (March 5, 1982): 1182.

has been well described elsewhere.[8] The assembly was essentially divided between those who wanted only as much outside interference with their work as was necessary to keep government officialdom (i.e., NIH and lawmakers) mollified, and those, among them the Berg Committee membership, who believed the risk potential considerable enough to warrant fairly elaborate research guidelines. As a series of votes on specific proposals eventually made evident, the latter group represented the consensus orientation.

The Asilomar dialogue has been characterized as pitting minimalists against moderates. The maximalists' position went unrepresented, it is contended, because few members of the scientific community favor formalized regulation of research.[9] But drawing ideological continua can be tricky business. For example, a key question running throughout the deliberations was how much self-regulation was necessary to fend off outside intervention. And, of course, a corollary question was what form this outside intervention would assume. The delegates were forced to address these matters directly when a battery of lawyers confronted them with a script of horrors that public policymakers might visit upon them. Did they know the extent of their liability in a civil suit, should one of their clones get loose and infect the countryside? Did they realize the wide scope of legislative discretion in providing public health standards? Regrettably, there were no constitutional lawyers on hand to discuss with them their free expression rights as scientific investigators working in the idea marketplace, no student of judicial politics who might tell them that the courts have a special responsibility to protect defensible interpretations of liberty against arbitrary legislative and bureaucratic action. That segment of opinion also went unarticulated in very great measure because, "by agreement, no one raised social or ethical questions."[10] "Legal" issues, yes; "social" issues, no. That constitutional questions transcended fish-fowl distinctions received scant attention. But more fundamentally, the assembly wanted to discuss only technical matters

8. Grobstein, *Double Image,* 23–28. For the assembly report, see "Summary Statement of the Asilomar Conference on Recombinant DNA Molecules, May 1975," reprinted in *Science* 188 (June 6, 1975): 991–94.

9. Grobstein, *Double Image,* 24–25.

10. Ibid., 25. Throughout the 1970s a small band of concerned scientists did make the necessary intellectual "connections"; unfortunately, they were well ahead of their time. The following serve as excellent examples: D. Stetten, "Freedom of Enquiry," *Genetics* 81 (1975): 415–25; D. Stetten, "Valedictory by the Chairman of the NIH Recombinant DNA Molecule Program Advisory Committee," *Gene* 3 (1978): 265–68; W. Szybalski, "Chairman's Introduction," in J. Morgan and W. J. Whelan, eds., *Recombinant DNA and Genetic Experimentation* (New York: Pergamon, 1979), 147–49; and W. Szybalski, "Much Ado about Recombinant DNA Regulations," in H. H. Fudenberg and V. L. Melnick, eds., *Biomedical Scientists and Public Policy* (New York: Plenum, 1978), 97–142.

amenable to rational accommodation. They assumed that their political mechanisms—debates, votes, etc.—were incapable of digesting social and ethical opinions. These dimensions were too "fluffy," or just too value-laden. But what could have been more amorphous than the issue of risk and hazard in the context of cloning? In fact, the whole question of what constituted a danger, and to whom, could be handled only through value predilections based upon speculations as to a wide variety of unknowns. The dialogue which ensued among the delegates purported to be an exercise in rational compromise over professional concerns; hence, the consensus could compliment itself on taking the middle-of-the-road position. But the test would come when the recommendations they approved became operational, putting empirically verifiable constraints on research.

The stage now shifted to the offices of the National Institutes of Health. Even before Asilomar, Paul Berg's committee had asked Director Donald Frederickson to establish an advisory panel which would promulgate suitable guidelines. Quite clearly, though, NIH was not going to announce anything until the cloners themselves presented a program of action. That blueprint took the form of a two-pronged laboratory containment system, the crux of the Asilomar consensus, and the specific details of which are discussed below. These physical and biological preventive modes the NIH Recombinant DNA Advisory Committee (RAC) juggled and embellished somewhat in close consultation with interested and influential gene splicers. The result was a series of administrative constraints which were to be binding on all NIH-funded cloning and which were submitted to the director as a set of recommendations for his final approval. With anxious researchers both here and abroad beginning to chafe under the moratorium, Frederickson held an open meeting of his own. Since RAC, like Asilomar, was dominated by professionals seeking professional answers, this session before the Director's Advisory Committee, a group charged with reviewing questions of public policy, constituted the last chance for a broader dialogue. Testimony for and against the RAC proposal was offered, but as long as the central issues remained technical, as in large measure they did, few could have expected significant change. Consistent with the politics of visibility, however, the director agreed to publish in the *Federal Register* his reasons for framing the restrictions as he saw them. The NIH guidelines, complete with annotated justifications, were issued on June 23, 1976. The self-imposed moratorium was over; federal government oversight and monitoring had commenced.

The first of the two sets of laboratory criteria established by the guidelines dealt with physical containment standards (P). To guard against the escape of potentially hostile organisms, the research environment must satisfy certain conditions, depending upon the level of experimentation. Thus, P1 (minimal) containment included the precautions commonly associated with sound research practice and was applicable to most genetic exchanges that could occur in nature among bacteria. P2 (low) containment included such precautions as keeping laboratory doors closed during experimentation and was applicable largely to genetic exchanges—typically involving DNA from either cold-blooded animals or plants and *E. coli*—that do not occur in nature but exhibit no evidence of hazard. P3 (moderate) containment included provision for an isolated work area where research would be conducted not on open benches but in Biological Safety Cabinets and was applicable to genetic exchanges between *E. coli* and various mammalian cells, birds, and embryonic primate tissue. Finally, P4 (high) containment included such exotic facilities as a special laboratory complete with airlocks and filters and was applicable to those genetic exchanges between animal viruses and *E. coli* where hazards seemed potentially serious.[11] The second set delineated criteria for biological containment (EK). To guard against dangerous consequences should a host organism escape, that organism (and, at that time, the only such guinea pig in use was the K-12 strain of *E. coli*) must satisfy certain minimal conditions of incapacity for harm, depending upon the level of experimentation. Thus, the least safe *E. coli* was the EK1 variety, which cannot inhabit the human bowel; the next safe *E. coli* was the EK2 variety, the chances for survival of which outside the laboratory were almost nil; while the most safe *E. coli* was of the EK3 variety and required such extensive testing that no samples were in use.

As these guidelines appeared in print,[12] they conveyed an air of supreme rationality. Not only were the rules pregnant with detail; not only did the two containment barometers evidence an ordered, incremental, sliding-scale effect; but they dovetailed handsomely. For instance, DNA extracted from plant viruses could be recombined into an *E. coli* plasmid under either P2-EK2 conditions or P3-EK1 conditions.

It is only when one consults the microbiological literature that the extreme severity of these rules begins to emerge. For example, the P2 constraints have long been considered sufficient for research involving such pathogenic bacteria as *Cholera vibrio*, while the P4 restrictions are the very same that prudent scientists would employ to study the most vir-

11. Stanley N. Cohen, "Recombinant DNA: Fact and Fiction," *Science* 195 (February 18, 1977): 656; Clifford Grobstein, "The Recombinant-DNA Debate," *Scientific American* 237 (1977): 33.

12. "Guidelines for Research Involving Recombinant DNA Molecules, June 1976," *Fed. Reg.* 41, no. 131 (July 1976): 27911–27922.

ulent microorganisms ever discovered in nature.[13] Moreover, the guidelines (with the unanimous approval of experts in the field) precluded federal support for any experiments employing genetic fragments from organisms *known* to generate hazard, such as the creation of bacterial recombinants containing the DNA of toxins and tumor viruses. So the containment criteria described above—and it cannot be emphasized too greatly that the entire concept of biological containment is unique to the gene-splicing context—were designed for those research enterprises where *no evidence* of risk had yet surfaced but where *no certainty* of risk-free consequences had been demonstrated.[14] Even some eminently sensible provisions proved unwieldy in the hands of Bethesda's bureaucracy. Thus, the guidelines made it impermissible for scientists under federal funding to release into the environment any recombinant DNA organism. NIH officials construed that rule to proscribe the growth of "manufactured" botanical specimens in any greenhouse or plot unless P2 conditions obtained, an interpretation plant geneticists viewed with alarm.[15] Finally, and of critical importance, many investigations allowed under the guidelines would have to await either the construction of P3 and P4 facilities or the development of EK2, not to mention EK3, artifacts. Eighteen months later, there were, at the most, only fifteen P3 facilities in the United States and no P4 installations at all engaged in cloning research.[16]

Certain specific supervisory turnstiles awaited recombinant DNA investigators subject to these standards. The parent institution was obliged to establish a biosafety committee, charged with overseeing relevant projects. No request for funding would be honored unless Bethesda approved both the grant application and the various physical and biological containment restrictions it considered appropriate to that proposal, but NIH officials stayed their hand until the particular biohazards panel cleared the way. Principal investigators were mandated to work up memoranda of understanding and agreements (the MUA) with these committees, specifying in detail the nature of their endeavors as well as mutually acceptable safety standards. The memoranda also included pledges that investigators would monitor their work in the light of those constraints, report laboratory violations, and submit for review, subject to both

13. Cohen, "Recombinant DNA," 656.

14. Ibid., 654.

15. P. R. Day, "Plant Genetics: Increasing Crop Yield," *Science* 197 (September 30, 1977): 1338.

16. U.S. Department of Health, Education, and Welfare, National Institutes of Health, *Recombinant DNA Research: Documents Relating to "NIH Guidelines for Research Involving Recombinant DNA Molecules,"* 3 (September 1978): 264, 401. Hereinafter referred to as *Recombinant DNA Research.*

home-base and national oversight, significant alterations in their research protocols. A key figure in the enforcement process was the biological safety officer (BSO), who examined laboratory conditions and recommended certification to the biosafety panel following a finding of compliance (see Fig. 3.1 for the twelve-step NIH flow chart describing this exercise in bureaucratic interplay). The campus committees also monitored—and insisted on MUAs regarding—experimentation conducted under the auspices of NSF, which had quickly adopted the entire guideline apparatus as its own except for the provision vesting the role of ultimate constraint-setter in the NIH study sections. Where a gene-splicing project was not federally sponsored but where the supervising institution required all staff members to abide by NIH criteria (as proved to be typical in the university setting), the above procedure applied except Bethesda's imprimatur. Finally, each MUA would have to be renegotiated on a year-to-year basis.

Neither NIH intervention in the campus research process nor the guiding hand of academic oversight committees was without precedent. Two years earlier, Congress had mandated all colleges and universities seeking federal funds in the name of their faculties for investigations employing human subjects to set up institutional review boards (IRBs). These committees were commanded, under applicable HEW guidelines, to review in advance all relevant research proposals and ascertain whether human subjects were placed at risk. If so, panels had to obtain assurances that "informed consent" had been tendered, and they also were to maintain surveillance over such projects so that the subjects' rights would not be compromised. Most provocative of all, however, was the discretion delegated the IRBs to balance the element of risk against the sum of the benefits which might accrue to participants plus the state of the research art. If subject risk outweighed these salutary goals, then experimentation could not proceed. Moreover, the IRB had special authority respecting research involving fetuses and pregnant women. Here it played a lead role in screening prospective subjects and monitoring the ways in which these participants were treated. Suggested interventions included oversight of the subject selection process and laboratory checkups of research in progress. Until an IRB report on stage-setting activities cleared Bethesda offices, no grant monies would issue.[17]

Comparing government policy on cloning research with government policy on human subjects research would seem to be unrewarding, if only because, *in theory*, public policymakers ought to enjoy greater discretion

17. 45 CFR, part 46, subparts A and B. Excluded from discussion are guidelines involving human *in vitro* fertilization, which probably cannot be justified as a free expression endeavor.

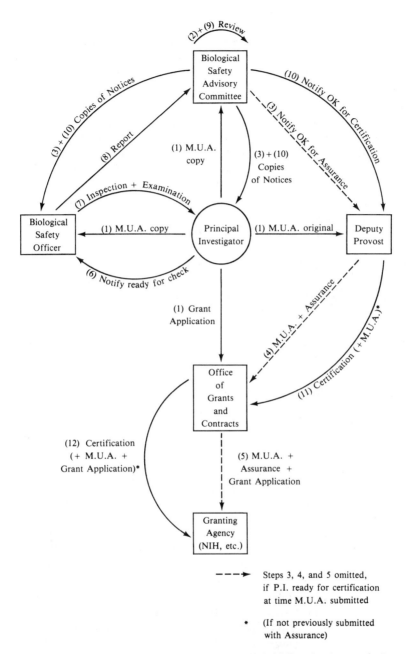

Figure 3.1. Procedures Mandated under 1976 NIH Guidelines for Approval of Cloning Experiments

Chart courtesy of Edward Adelberg, Chairman, Yale University Biosafety Committee

in protecting human *subjects* than in protecting humans who may never confront a cloned organism. To that extent, the greater flexibility and, hence, the greater power seemingly accorded the IRB as compared to the biosafety committee rings reasonable; on the other hand, there have never been federal guidelines relevant to human experimentation research which remotely compare either in detail or in rigor with those advanced in 1976 for recombinant DNA investigations. But, of course, the central wisdom which emerges from the preceding analysis is that theoretical estimations are not controlling. It can be no coincidence that these twin sets of guidelines and oversight panels were launched within the same short span of time, later to be followed by strictures governing abortion and recent controversies surrounding NIH enforcement of constraints on the use of monkeys as research subjects. Unlike the Framers, we perhaps do not recognize the king's licensor when we see him. Today, that licensor is the federal government's funding agencies, turning apples into oranges, treating them all of a piece because these are "provocative behaviors," and it is the duty of the "people's bureaucrats" to protect the citizenry from potential excesses which might arise from the new, the unusual, and the ethically suspect, especially where the citizens' monies are involved. That these constraints were aimed, in part, against constitutionally protected science was an argument that apparently never occurred to the regulators, just as the constitutional implications of a free press never occurred to the Crown's deputies. But many a recombinant DNA researcher might have said in rebuttal, "If NIH doesn't act, others will, and even more aggressively." That story we now address.

As major cloning efforts became slated for action at the campus level, the recombinant DNA debate, naturally enough, took up residence at several important research centers, such as Cambridge and Ann Arbor. In these precincts could be found not only a broad representation of concerned citizens and scholars but also the remnants of the antiwar, counterculture coalition, perhaps ready at this time to shift gears from a discussion of the alleged political self-interest of the governmental establishment to a dialogue on the alleged research self-interest of the scientific establishment. An unusual feature of the Cambridge episode, which became by far the most visible and significant of the local imbroglios, was that organized opposition to genetic engineering centered not on some new student personality of the Tom Hayden-Mario Savio species but on the mayor of the city, an establishment figure if ever there was one. But then that development was no surprise at all to those familiar with Greater Boston politics.[18]

18. Again, I rely on Grobstein's fine descriptive account. See chap. 5 of *A Double Image*. I do not mean to imply that the Cambridge commotion was typical of local reac-

Cambridge, Massachusetts, is a curious social mix, comprising, on the one hand, a citadel of academic excellence and a prime example of what C. Wright Mills would term "power elitism," Harvard University, and, on the other hand, a surrounding blue-collar Caucasian population hardly attuned to the ways of either the Ivy League or of social upheaval. Few communities have suffered more the travails of town-gown friction; and Alfred E. Vellucci, the city's chief executive officer, well understood how to maximize these tensions for political capital, championing the interests of his constituency, John Q. Cambridgian, against the perceived "enclave of privilege and nonconformism." When Harvard's gene splicers sought to develop P3 facilities and received permission from university authorities to proceed, Mayor Vellucci stood ready to intervene.

Not that Mr. Vellucci was the chief architect of outcry. He and, later on, the Cambridge City Council were political lightning rods, attracting the attention and hopeful cooptation of dissenters in the Harvard recombinant DNA dispute. The activities of these pressure groups deserve specification. Taking the lead among well-known scientists were Nobelist George Wald and Ruth Hubbard, both Harvard biologists, who visited Vellucci and asked him for support. Wald considered the safety issues secondary, arguing that this research should not be done at all because it was a corruption of nature.[19] Also making a direct overture to Vellucci was Francine Simring, who represented the "genetics watchdog committee" of Friends of the Earth, a prominent environmental group. She appealed for a moratorium on all cloning until relevant questions of health and ethics were resolved. Other protesters drew a direct parallel between nuclear and recombinant DNA research, citing them as twin horrors capable of mass societal harm. Finally, there is to be noted the attitudes of Science for the People, a Cambridge-based group of younger activist professionals. In their view, cloning was the product of elitist-oriented practitioners with a vested concern for their own accomplishments; in no way did the national interest or the public interest justify these probings. To them, "big science" had become too closely aligned with the technocratic values of "big government," and genetic research as a whole was placing undue emphasis on the role of natural phenomena in contradistinction to the role of inequitable social norms and power structures as mechanisms ordering human behavior. Fortified by so impressive an array of endorsement, Vellucci took the offensive, announcing that the

---

tion. In some communities, notably Madison, recombinant DNA research received a clear show of support, while in others, such as Princeton, ordinances were enacted with considerably less fanfare.

19. National Academy of Sciences, *Research with Recombinant DNA: An Academy Forum* (Washington, D.C.: NAS, 1977), 55.

health and safety of Cambridgians must be protected against the threat of diseases and monsters, and recommending that gene-splicing research be banned *en bloc* within local boundaries. The City Council's reaction to all this seemed measured and temperate: if Harvard and MIT would put a three-month moratorium on their recombinant DNA experiments, a special blue-ribbon citizens' panel would investigate the matter and issue a report. The council could then act in responsible fashion. Needless to say, the two universities complied.

Why anyone expected the Cambridge Experimentation Review Board (CERB) to make a telling contribution to the recombinant DNA debate is not clear. I have argued that decisonmaking at both Asilomar and Bethesda tended toward a narrowness of focus, though significant political and social questions were ripe for dialogue. This void CERB might have sought to fill, but it disclaimed any such responsibility. The City Council's charge related only to public health, not to "philosophical issues," its report announced.[20] But if Cambridge were going to construct merely a mini-NIH set of risk estimates, was it plausible to believe a "citizens jury," as some members called themselves, could improve on Bethesda's guidelines? To this, the report gave the ambiguous statement: "Our role was to examine the controversy within science."[21] If the debate to be addressed was one of safety considerations, then the evidence collected and the judgments rendered must also relate to safety, the very issues upon which NIH had already labored. We may conclude that what both sides hoped for from CERB was not informative policymaking but "half a political loaf." Mayor Vellucci's forces wanted to slow the march of cloning research; Harvard and MIT wanted to blunt Vellucci's sallies. If that is what these antagonists desired, they succeeded.

CERB submitted its recommendations in January 1977, six months after commencing activities, during which period the self-imposed moratorium was extended. The panel's major finding was that P3 research could go forward, but that the NIH guidelines were insufficiently protective. One scans the CERB report in vain for reasons for the added measures. The committee did congratulate itself on reaching a unanimous judgment despite its cross-sectional makeup, but even a petit jury has the benefit of legal standards in the form of the "beyond a reasonable doubt" rule when it debates verdict options of guilty or not guilty in criminal cases. What legal norms informed CERB? Was the burden of proof on the recombinant DNA scientist or on the community? And what was it that one or the other must demonstrate? These questions the

20. "Guidelines for the Use of Recombinant DNA Molecule Technology in the City of Cambridge, January 1977," reprinted as appendix 4 in Grobstein, *Double Image,* 152.
21. Ibid., 153.

group sloughed off as follows: "There was no clear consensus on the issue of who must justify what."[22] CERB had been set adrift upon a boundless legal and constitutional ocean, whereon only visceral reactions and political expediency were likely to control.

Judged by the standards of scientific usage and constitutional principle, CERB's added assurances were hardly defensible. Three reservations deserve special mention. First, CERB asked that P3 experiments use EK2 biological containment; EK1 organisms as host vectors did not suffice. At that time, only a very few EK2 systems had been found bona fide according to RAC technical assessments; certification for others might take years. So, in reality, P3 work could proceed only on the most limited basis. But to focus on the question of expertise, for local governance to disallow by law P3-EK1 cloning, hence challenging directly NIH health and safety policy, while constitutionally permissible, borders on audacity. Second, CERB determined that the Harvard and MIT biosafety committees must include one community representative each, individuals approved by the Cambridge Health Policy Board. By what "supremacy of law" theory does a city instruct a federal agency how its committees should be constituted and how they should meet their responsibilities? John Marshall told us long ago that national action, where permissible, preempted conflicting state action.[23] There was not then, nor is there today, any local option privilege attached to the NIH guidelines. And by what standard of living constitutional decorum does a city instruct private institutions of higher education resident therein that they must include local policymakers on their committees? CERB's recommendation is virtually impossible to reconcile with traditional notions of academic freedom.

Third, Cambridge was advised to organize its own biosafety panel which would prescreen all cloning research to make sure it comported with NIH constraints and to undertake inspections of campus installations. I have pointed out that communities possess ample power to regulate the "times, places, and manner" in which scientific experimentation is performed. Without doubt a city can establish biohazards agencies and conduct on-site reviews of workplaces. This does not, however, give local government carte blanche to censor research in the nature of quasi speech for the purpose of ascertaining that its content squares with what the federal government allows. The stringency which characterized the NIH 1976 guidelines has been duly noted above, but of course, stringency is not synonymous with impermissibility. Rulemakers can insist upon, in the form of funding strings, reasonable criteria as to how ex-

22. Ibid., 156.
23. Gibbons v. Ogden, 9 Wheat. 1 (1824).

pressive activity shall be conducted, and the NIH physical and biological containment standards seemingly satisfied this test. But NIH typically is subsidizing the experimentation at issue; Cambridge has no financial or other resources involved in these research enterprises. Of course, the city can take NIH's restrictions for its own, enforcing them, say, through the medium of the criminal ordinance. But if it is going to set up a recombinant DNA censorship board, it will have to satisfy relevant constitutional doctrine. The Supreme Court has said that cities can license parades— a form of protected expression—but it has never said that a city can examine in advance the content of the parade message. The Supreme Court has stated further that cities can license motion pictures—also a de facto form of quasi speech as I have argued elsewhere[24]—but only to quarantine obscenity, i.e., "nonspeech," and only under such strict procedural protections as timeliness, ready access to judicial forums for enforcement purposes, and a burden of proof calculus which the community must shoulder.[25] Absent these guarantees of fairness in the censorship process, this version of king's licensure would appear to fail.

On February 7, 1977, the Cambridge City Council unanimously adopted the CERB recommendations. Before doing so, however, it delivered the coup de grace to the indefatigable Mr. Vellucci, rejecting by a 6 to 3 vote his motion to ban all forms of gene splicing within the city's jurisdiction.[26] His proposal, the most extreme ever maintained by a notable public official, could be entertained as constitutionally viable only if one were prepared to argue that (*a*) cloning is all conduct and no expression, or (*b*) cloning, like obscenity and "fighting" words, may be proscribed totally, *and on the basis solely of its content,* because in all its species and forms, the methodology either lacks redeeming social value or constitutes a clear and present danger to the community. As I contended in Chapter 2, the former proposition places the scientist entirely at the mercy of duly constituted authority, while the latter contention is without any factual basis. Two and one-half years after the Berg Committee's initial call for a moratorium on "sensitive" experiments, Cambridge academicians could now perform at least some recombinant DNA research.

That CERB-type bargaining was not going to provide a long-term political accommodation between competing value interests was made

24. Ira H. Carmen, *Power and Balance* (New York: Harcourt Brace Jovanovich, 1978), 231-33.

25. Freedman v. Maryland, 380 U.S. 51 (1965).

26. Nicholas Wade, "DNA: Laws, Patents, and a Proselyte," *Science* 195 (February 25, 1977): 762. An update on the Cambridge scene is provided in Sheldon Krimsky, "Local Monitoring of Biotechnology: The Second Wave of Recombinant DNA Laws," *Recombinant DNA Technical Journal* 5 (June 1982): 79-84.

evident less than a month later. The National Academy of Sciences had decided to sponsor a forum in the nation's capital to dramatize, for the edification of a watchful political audience, the openness of the recombinant DNA research community to a frank exchange of views.[27] Militant opposition forces saw this gambit as a fraud: NAS, a quasi-public agency basking in the status of its esteemed membership, was lobbying the power structure for greater funds and lesser constraints in the name of clientele scientists; and these specialists would repay the debt by manipulating this exotic new tool to solve many of the nation's most pressing social problems. But those prepared to question the cloning methodology were not themselves an ideological monolith: they ran the gamut from informed, constructive skeptics to "falsetto screamers," a fact that became clear when the program planners, realizing there would be retribution in the form of demonstration and disturbance if they did otherwise, provided the outsiders with "unequal time" to express their views. For example, one environmental spokesman emphasized that industry had begun to develop recombinant DNA projects while remaining free to adopt or reject in piecemeal fashion the government restrictions binding on academicians. Fair enough, for here was an issue which Congress could confront in a manner consistent with both its control of interstate commerce and free expression parameters. Jonathan King, an MIT biologist, came down hard on Asilomar, calling its treatment of cloning issues one-sided. Also fair enough. There was indeed a whole host of other dimensions deserving careful deliberation which ought to be permitted to surface in a marketplace capable of assimilating and assessing disparate outlooks. NAS, in its own perception of "balanced treatment," had included non-cloner biologists on the speaker list: Robert Sinsheimer, Jonathan Beckwith, Erwin Chargaff. But nowhere could be found the theologians, the legal scholars, the social scientists, and the ecologists. On the other hand, the audience could hardly respect Jeremy Rifkin of the People's Business Commission, whose strident, "Let's open this conference up [to the people] or close it down!" brought back memories of such anti-establishment lawlessness as the Mark Rudd, Columbia University debacle. Also represented at the cacophonous end were the unceasing cries of "moratorium," this time emanating from some members of a new group called the Coalition for Responsible Genetic Research, an umbrella organization comprising Science for the People, the Environmental Defense Fund, Friends of the Earth, and concerned critics such as George Wald and Lewis Mumford.[28]

27. Grobstein, *Double Image,* 72.
28. Ibid., 73. The opinions cited in this paragraph are drawn from remarks found in NAS, *Research with Recombinant DNA,* 18–21, 38–39, 222–23.

Needless to say, NAS owed no duty whatsoever to these uninviteds, including self-appointed representatives of *vox populi,* and it may fairly be asked just what a ban on P1 and P2 cloning would accomplish unless it were believed, as few did believe, that gene-splicing research was a prelude to Huxleyan disaster and, hence, inherently wrong.

An important underlying question lurking throughout the speechmaking was how best to characterize the attitudes and values of recombinant DNA scientists toward political accountability as an alternative to political freedom. Two divergent profiles were submitted. The first, coming from Maxine Singer, an NIH scientist and Asilomar activist, depicted investigators as attentive to public issues, well prepared intellectually to weigh the right of inquiry against the common good. The second, submitted by Jonathan King, painted a picture of grasping, asocial egoists, implying that they marched to the tune of Nobelism and the rhythm of technocracy. Each cited Asilomar and its progeny of self-regulation as proving the case presented.[29] But, of course, Asilomar proved neither. It was merely a practical, cautious stopgap offered in the face of trepidation, confusion, and the lack of larger focus, signifying little of importance about basic drives and ideologies.

Certainly no one could say that interested scientists, as they floated out new regulatory blueprints, were yet prepared to look for suitable accommodations through the medium of constitutional principle. Take, for instance, Professor Sinsheimer's suggestion that cloning be confined by law to a single P4 facility under the control of the federal government.[30] This notion, of course, would preclude university-based research and would put public officials into the business of deciding who could participate and under what terms. It is difficult to conceive that the First Amendment permits the nation's political leadership—absent a compelling interest such as the national defense—to create a monopoly for the benefit of those investigators and those aspects of expressive activity considered most deserving. Sinsheimer's argument appears to have been predicated on the analogy, articulated both in media reports and through various critical forums, that cloning lies on the same scientific (and ethical) plane as atomic bomb investigations. Regarding the scientific comparison, the analogy is patently spurious. Production of nuclear weaponry may properly be considered comparable to the harnessing of recombinant DNA science for purposes of biological warfare. The United States is a party to the Biological Weapons Convention; and it has interpreted this legal obligation as barring the development of clones

29. NAS, *Research with Recombinant DNA,* 24–30, 38–40.
30. *Recombinant DNA Research* 1 (August 1976): 438.

for offensive, though not defensive, military ends.[31] To be sure, the federal government exercises control over *all* atomic research, but as was noted in Chapter 2, until such time as one can speak realistically about recombinant DNA diseases and monsters, science fiction scenarios should not be permitted to legitimate public policy paradigms, where the marketplace of ideas is a salient consideration. Another suggestion was that *E. coli* should be declared off limits as a host for cloning experiments because that bacillus prefers the human intestinal tract. Experience indicated, even then, that *E. coli,* and especially the K-12 strain, is an extraordinarily adaptive instrument for gene-splicing procedures, some of which may conceivably pose hazards but others of which are eminently safe. Again, the proposal paints with too broad a brush and smacks of banning cyclotrons, telescopes, printing presses, and other paraphernalia considered at one time deleterious but without which expressive activity as a significant attribute of our intellectual/constitutional heritage might well be straitjacketed.[32]

But the most significant—and ambitious—policy proposals came from the nation's political leaders. As a result of the incessant press coverage devoted to the subject during early 1977 plus the finding by a special executive department committee that no single federal agency currently possessed the legal authority to monitor all aspects of ongoing cloning research,[33] both President Carter and Senator Edward Kennedy, his leading challenger for majority party leadership in the Congress, put forward a package of comprehensive legislative regulations. Mr. Carter's plan was unveiled in April 1977. It would have instructed the secretary of HEW, Joseph Califano, to promulgate as interim standards applicable to *all* recombinant DNA research the NIH guidelines currently binding upon only federally subsidized studies. One year later the secretary was to announce permanent criteria. The proposal also gave him sweeping authority to license gene-splicing facilities and register gene-splicing projects, all of which would have to meet the appropriate guidelines. With regard to these various determinations, the bill delineated absolutely no standards whatsoever to fetter the secretary's discretion.[34] The Kennedy measure was reported out of committee in July 1977 by a vote of 13 to 1 and constituted a revision of the Carter approach.[35] There, the licensing,

31. Charles Pillar, "DNA—Key to Biological Warfare?" *The Nation,* December 10, 1983, 596–601.

32. Nicholas Wade, "Recombinant DNA: New York State Ponders Action to Control Research," *Science* 194 (November 12, 1977): 705.

33. Grobstein, *Double Image,* 160.

34. S. 1217, 95th Cong., 1st sess., 95 *Cong. Rec.* 5335 (1977).

35. S. 1217, Calender No. 334, 95th Cong., 1st sess. (1977).

registration, and guideline-formulating tasks would reside in an independent forum to be called the National Recombinant DNA Safety Regulation Commission, the majority membership of which would not be biological investigators. Reacting against the paucity of legislative standards in Mr. Carter's recommendation, the Kennedy bill required that installation licensing be conditioned upon whether the commission believed experiments conducted there were pursued "in a manner as to protect the health of the persons exposed to recombinant DNA, protect the environment, and protect the health of the population of the surrounding community." It also charged the commission to develop precautions "no less stringent than" the NIH guidelines "for purposes of protecting the health and safety of individuals who work with recombinant DNA, the health and safety of the public at large, and the integrity of the environment." Moreover, those who asked for permits must agree to allow government inspectors on their premises, and those inspectors not only might peruse relevant equipment, files, and papers but also might conduct such searches without warrants. Finally, the commission could approve state and local regulatory schemes more stringent than the federal constraints, should they be "relevant and material" to such "compelling" interests as health and "environmental concerns," and if they were not "arbitrary and capricious."

As these efforts represent the most considered attention that the president and the Congress have to this writing (1985) accorded the recombinant DNA debate, they deserve intensive constitutional scrutiny. From the standpoint of constitutional *law*, the overriding flaw with both packages is that they eschewed any conception of cloning research as implicating the higher virtues of scientific inquiry and, consequently, the higher virtues of free expression. Time after time, the Supreme Court has struck down censorship programs as being especially condemnable under First Amendment standards; as I urged relative to the Cambridge licensing statute, only in the atypical instances wherein government has sought to isolate obscenity[36] and ensure that parades conform to traffic exigencies[37] have the justices countenanced prior restraints as a matter of course.[38] Yet here were statutory blueprints that would have placed every institution of higher learning and every distinguished professor engaged in recombinant DNA investigations at the mercy of the federal bureaucracy, functionaries of which could henceforward leaf through

36. Times Film Corp. v. Chicago, 365 U.S. 43 (1961); Freedman v. Maryland, 380 U.S. 51 (1965); United States v. Thirty-Seven Photographs, 402 U.S. 363 (1971).

37. Cox v. New Hampshire, 312 U.S. 569 (1941).

38. For a complete analysis of the doubtful constitutionality of censorship as a regulatory tool, see Ira H. Carmen, *Movies, Censorship, and the Law* (Ann Arbor: University of Michigan Press, 1966).

research proposals and facility qualifications, deciding which were consistent, not merely with NIH guidelines as per Cambridge, but with *their own* conceptions of health and safety, and which were not. It is perhaps worth noting that when Birmingham, Alabama, declined to license a civil rights parade—which, like the San Diego experiments discussed in Chapter 2, is an example of quasi speech—on the ground that it was inconsistent with "health and safety" standards, the Supreme Court reversed, finding the ordinance void on its face.[39] With respect to the "search and seizure" question, the notion that scientists must submit to warrantless intrusions because of the *content* of their inquiries is unprecedented. Finally, as I have before emphasized, recombinant DNA endeavors are just as likely to be expressive activity and harmless as they are to be commercial ventures and harmful. Indeed, even "commercial speech" is entitled to some—though not a full measure of—First Amendment consideration.[40] State and local gene-splicing regulations applicable to the university laboratory environment must carry a far higher burden of proof than merely to qualify as "relevant" and "reasonable" implementations of traditional Tenth Amendment prerogatives, unless they are specifically limited to "times, places, and manner" contingencies. That public officials may choose to label their conventional interests as "compelling" hardly makes them so when these are weighted against basic freedoms.

From the standpoint of constitutional *politics,* these provisions can only be understood fully against the backdrop of prevailing liberal ideology. As Lowi points out,[41] the federal law-making process during much of the post–World War II period has been a creature of "interest-group liberalism," the term he employs to describe the public philosophy of the day. His argument is that while liberals favor central governmental initiatives respecting social policy in the grand FDR tradition, their Jeffersonian sensibilities (absent national emergency) impel them to hedge their bets against the coercive force which power allocations will perforce accomplish when articulated through coherent, carefully drafted regulatory standards. The result of this Janus-faced attitude toward national political responsibility has been a series of federal statutes which delegate to key interests the authority to govern themselves via the medium of purposely vague legislative criteria. Roosevelt himself presumed in favor of "constitutional pluralism" when he mounted his ill-fated Blue Eagle program. Certainly if the theory is descriptive of "political branch" activism in Lyndon Johnson's day, a contention whose merits Lowi

39. Shuttlesworth v. Birmingham, 394 U.S. 147 (1969).
40. Virginia State Board v. Virginia Citizens Consumer Council, 425 U.S. 748 (1976).
41. Theodore J. Lowi, *The End of Liberalism* (Chicago: Norton, 1969).

demonstrates with considerable skill, then there is no reason to think it should have paled during Jimmy Carter's tenure. In fact, the evidence is considerable that "interest-group liberalism" was alive and well during the 1970s, as a reading of the Economic Stabilization and Clean Air Acts, among other legislative measures, makes manifest.[42]

One does not even have to know that President Carter's proposal was drafted by a Federal Interagency Committee chaired by NIH Director Frederickson to see that it fits the Lowi model practically to a *T*. The secretary of HEW was given a blank legal check to contrive and enforce such regulatory criteria as he deemed necessary. But NIH was part of his shop, and the NIH director was his subordinate in the bureaucratic chain of command. It is inconceivable that he would have gutted his own department's guidelines; at most they would have been amended in the light of new evidence. But as we know, the guidelines themselves were based on the Asilomar consensus; these, according to NIH policy, could be altered only as RAC proposed and the director ratified new standards. And RAC was and is controlled largely by the molecular biological community. So what the Carter legislative package would have accomplished was as close to interest-group self-government as political exigencies permitted.

On the surface, the Kennedy approach seems conceptually different, because an "independent" commission of "outsiders" would make the most important decisions. That sort of policy-building apparatus smacks of La Follette Progressivism, which has given us the ICC, the FTC, and other regulatory agencies, allegedly insulated against conflict-of-interest cooptation. The political science literature, however, holds out few illusions on this score; that is, because these agencies are cut off from executive and legislative moorings, they become easy pickings for their "constituencies."[43] And the Kennedy bill invokes the all-encompassing "health and safety" criterion almost as frequently as does the Clean Air Act's authorization to the Environmental Protection Agency. If history were to be our guide, a National Recombinant DNA Safety Regulation Commission would probably have done little more than legitimate what "establishment" natural science would have approved. Not that the senator had planned it that way, of course; but the unstructured regulatory norms to which he gave his support showed he was not ideo-

42. Cf. PL 91-379, Title II (1970) with PL 91-604 (1970). For further analysis, see Carmen, *Power and Balance,* chap. 4.

43. Excellent accounts of this phenomenon are found in Marver Bernstein, *Regulating Business by Independent Commission* (Princeton: Princeton University Press, 1955); Samuel P. Huntington, "The Marasmus of the I.C.C.: The Commission, the Railroads, and the Public Interest," *Yale Law Journal* 61 (1952): 467–509. For broader discussion, see Carmen, *Power and Balance,* chap. 7.

logically prepared to make rational, systematic, predictable oversight through law a reality.

More difficult to square with "interest-group liberalism" is the local option feature of the plan. Recombinant DNA scientists had had their fingers burned once already at Cambridge. If they feared a national commission officially removed from NIH, then they were horrified at the prospect of interminable brushfire skirmishes with the local activists of Princeton, Palo Alto, Berkeley, and Baltimore.[44] Halting and uncertain though this section of the bill was—a grassroots community could not proceed except under federal approval—it did represent a rather different kind of public control. The provision smacked of "moral uplift" politics: it embodied the strain of liberalism which informed Warren Court policy thrusts in the 1960s; it also constituted one instance among many of Mr. Kennedy's *modus vivendi* with Ralph Nader–Common Cause–Sierra Club "Mugwumpism." What seems curious about the notion is that "new-style liberalism," as this value orientation has been called,[45] emphasizes strongly the element of national problem-solving. But of course, according to the Kennedy blueprint there could be no opportunity for the people "back home" to do anything but *toughen* federal ground rules, so we do not have here any bouquet being tossed in the direction of a viable states' rights doctrine.

It is rather extraordinary that these, and other, legislative initiatives should have lost their momentum in a matter of only a few months. True, extensive hearings were held by as many as four congressional committees at this time, during which all interested parties had a chance to express their views,[46] but it would be a grand leap of faith to conclude that the lawmakers eventually reached consensus on the significant issues of the moment and decided that public regulation by statute was simply inappropriate. Rather, the demise of the Carter-Kennedy packages can be understood only through an appreciation of rumblings deep within the scientific community. These developments show that molecular biologists, for better or worse, were no more prepared to address the larger questions posed by cloning at this time than they had been at Asilomar.

First, many researchers who had gone along with the NIH guidelines in

44. Grobstein, *Double Image,* 76.

45. Carmen, *Power and Balance,* 367–78.

46. E.g., *Regulation of Recombinant DNA Research,* Hearings before the Subcommittee on Science, Technology and Space of the Committee on Commerce, Science and Transportation, U.S. Senate, 95th Cong. (Washington, D.C.: Government Printing Office, 1978); *Science Policy Implications of DNA Recombinant Molecule Research,* Hearings before the Subcommittee on Science, Research, and Technology of the Committee on Science and Technology, U.S. House of Representatives, 95th Cong. (Washington, D.C.: Government Printing Office, 1977).

great measure because they considered them the least of all possible evils once the Pandora's box of scientific tremulousness had been opened were now fed up with the entire censorship regimen. To this effect was James D. Watson's mocking assertion, intentionally directed, through the pages of the *New Republic,* toward the intellectual liberal establishment: "I find almost universal agreement among leading molecular biologists that (our miserable) guidelines are a total farce."[47]

Second, several practitioners had been devoting their full attention to biosafety matters, and armed with such new EK2 strains as the chi-1776 host and pBR322 vector, they were now prepared to argue that their host vector systems not only lacked survival capacity outside the laboratory but also eliminated virtually all potential hazard, unless, of course, the investigator cared to recombine a gene for some toxic substance and then swallow the clone.[48] Moreover, there was now a good deal of evidence showing that genetic exchanges among diverse bacteria in nature were far more frequent than had previously been understood, while new risk assessment studies had demonstrated that viral DNA could be cloned safely in *E. coli* and that crippled viruses could be used as vectors for cloning DNA materials extracted from eukaryotes.

Third, the silent majority among interested biologists had become thoroughly disenchanted with the prospect of congressional intervention and decided it was high time to enter the lobbying lists. Participants in several Gordon Conferences sent petitions to every member of Congress in which they attacked the threat of unreasonable constraints on scientific investigations. Then, under the skillful leadership of Harlyn Halvorson, the president of the American Society of Microbiologists, a coalition of research institutions sent a strong letter to the House and Senate; deputations representing such prestigious organizations as the American Association for the Advancement of Science were also dispatched to congressional offices sounding the word that, in the light of new evidence, the entire issue of biohazard had been greatly overblown. After all, these scholars asked, had one single episode of actual harm occurred as a result of splicing genes?[49] In the face of this rising tide, Senator Kennedy withdrew his bill, and President Carter shelved his recommendations. The Congress has yet to vote on any policy aspect of the recombinant DNA debate.

We are now ready to address the argument posed several pages earlier, namely, that the NIH-Asilomar framework could be justified as a prophylactic strategy, deterring more assertive regulations by nonscientists. It is likely that, without some set of constraints, voluntary or otherwise,

47. James D. Watson, "In Defense of DNA," *New Republic,* June 25, 1977, 13–14.
48. Grobstein, *Double Image,* 89–90.
49. Ibid., 80–81.

the recombinant DNA research community would have encountered much tougher sledding before suspicious city councilmen, state legislators, and congressmen. But the question recurs: *What set?* And another question recurs as well: *Under what hierarchy of principles ought a set of guidelines be built?*

It is beyond dispute that with rare exceptions neither NIH officials nor Asilomar delegates were attuned, for whatever reason, to discuss scientific inquiry and gene splicing as constitutionally relevant political phenomena. But that does not mean that implicit conceptions of constitutional value went unrepresented in decisionmaking forums. The root concept brandished in support of the prevailing modus operandi was and is "public consent." One analysis contends that because the prime purpose of free expression is to further self-government, enlightened republicanism includes the role of law as a kind of informed approval, permissibly setting forth the do's and don'ts of any behavior—including cloning—which could alter the nature of the human condition.[50] Another commentator observes that NIH tacitly analogizes between the person as consenting agent in the human subjects research context and the public as consenting agent in the recombinant DNA research context. With the former, the state insists on receiving proof of informed permission before experimentation can proceed; with the latter, the state represents the voice of the people and insists on the right to veto investigations considered inappropriate.[51]

Both arguments appear dubious on their face. It strains credulity to believe that when microbiologists recombine DNA materials in their university laboratories for the purpose of understanding the nature of genetic properties, the public assumes the status of "subject," vested with the police power to withhold consent regarding each and every phase of the experimental process. But more basic still, in the American scheme of governance, the central theme is not public consent; it is ordered liberty. This essay sees the marketplace of ideas as belonging to all persons, not in the sense that a majority, acting in the name of some "general will," can stop everyone else from inquiring into the condition of the species, but in the sense that humankind is free even to denude intellectually its own reality, as long as the exploration is but a search for truth and nothing more. And when that search for truth does encompass something more than "speech," governmental strictures must be carefully tailored, balancing with a reasoned eye the scientist's zone of free investigation and the public's zone of social duty.

Perhaps the most glaring deficiency in the entire recombinant DNA

50. Marc Lappé and Patricia Archbold Martin, "The Place of the Public in the Conduct of Science," *Southern California Law Review* 51 (1978): 1535–54.

51. Grobstein, *Double Image,* 93.

debate even to this writing is that virtually all participants have failed to articulate and apply accepted standards of constitutional value to the unfolding myriad of circumstances. Who is at fault, if anyone? The scientists? They were under the gun at Asilomar. Later, when, understandably, they became alarmed at the fates lawmakers had in store for them, they took a "that government is best which governs least" attitude. NIH bureaucrats? Their job is to authorize money plus whatever strings public pressure imposes on clientele interests. The politicians? In this day and age, they gravitate between the notion of science and technology as constituting a slippery slope of concern and the notion of science as responsible discipline, its excesses susceptible to practitioner self-governance under the umbrella of bureaucratic oversight. Lawyers? They represent clients, and the courts have not yet been called upon to confront cloning. Political scientists? They like to think of themselves as "doing" science, but they have rarely studied science as a constitutionally relevant phenomenon. In sum, there is no blame to pass around in the form of personal culpability or negligence. There exists, however, a systemic problem, a self-made dilemma involving political theory, intellectual understanding, and social reality. Relationships between the individual and the state, the Founding Fathers believed, must be governed by rules. At their best, these rules are not merely givens handed down either by force of habit or by decree. They are discussed, explained, understood, applied, and, where need be, refined. Nor can they be neutral, for they give to some and withhold from others, depending upon the particular blend of social values necessitating accommodation. That knowledgeable people tend not to see the political process as the people's mechanism for explicating coherent, comprehensible, and vital norm structures, applicable to all significant power allocations between researcher and public, is really a confession that we do not permit the Constitution to perform its assigned office. That is why the recombinant DNA debate has sometimes been a symphony written in an unknown key. But I have described only the first movement.

Elated at their triumph before the bar of congressional opinion, recombinant DNA scientist-activists now turned their fire on the barely dry guidelines. The NIH director's expert committee on cloning procedures (RAC) got the ball rolling even as the Kennedy and Carter bills were grabbing center stage; its recommended revisions were announced in the *Federal Register* on September 27, 1977. Three months later, the Director's Advisory Committee held a two-day public meeting on the proposed changes. As to safety issues, those who thought the impetus toward deregulation was escalating too rapidly were clearly on the defensive. The United States had made the big theoretical breakthroughs in the new discipline, but European specialists were performing experiments that could not be done here. Robert Sinsheimer was now ready to exempt all P1-EK1 research from the guidelines, a recommendation which even went beyond RAC's amendments.[52]

But there was also much discussion about nonscientific matters, as befitted this particular forum. Leslie Dach of the Environmental Defense Fund labeled the restrictions vague in the extreme, incapable of objective implementation. Clearly, the RAC membership was badly in need of people with legal/administrative skills, he submitted. And how could that committee formulate rules for the public interest when only two of its fourteen representatives were outside the scientific community? To this, the NIH director, Dr. Frederickson, reminded the forum that NIH was reformulating *guidelines,* not *regulations;* in fact, he believed it was high time for Bethesda to give the institutional biosafety committees (IBCs) much more leeway. If people wanted laws, they should go to Congress, which NIH had already done, the director noted. Patricia King, a member of the National Commission for the Protection of Human Subjects, replied that NIH had never been quite able to decide whether it wanted to be an advisory or regulatory body. All in all, she preferred the former role, because sponsoring and governing research was a *mésalliance,* as the AEC's demise had shown. In fact, however, the NIH rules went beyond guideline status, for these contained all manner of specifications regarding the responsibilities of IBCs, administrative supervision which was really the business of a much more representative panel modeled after, say, the one on which she served.[53]

But RAC received a few knocks from the other side as well, from those who thought the panel was parading as science not only law but also politics. One commentator challenged a committee pronouncement that the revisions had been made intentionally restrictive so as to err on the side of caution. RAC's job, he said, was to provide professional sophistication, not political accommodation in the form of "excessive or unsound *regulations*" (italics mine).[54] Another speaker, James D. Watson, said that legal rules—whether bureaucratic strings attached to gifts or statutory enactments—should be designed to control human behavior only when there was evidence of danger. Nowhere did RAC purport to make such a case; hence, it should get out of the guideline business without even considering political amenities.[55]

Evidently sensing that mainstream opinion favored a mix of modest

52. *Recombinant DNA Research* 3 (September 1978): 278.
53. Ibid., 232–34, 482–83.
54. Ibid., 379.
55. Ibid., 438, 440, 446.

rule-softening and modest decentralization, Frederickson barely broke stride when Senator Adlai Stevenson (D-Ill.) asked him for an official explanation as to why NIH had been tardy in investigating a breach of the rules at the University of California in San Francisco. There, the research team of Howard Goodman, Herbert Boyer and others had employed a RAC-approved but still uncertified EK2 vector, a pBR322 plasmid. The fact was, said the NIH director, that primary investigative responsibilities devolved upon the local IBC, and that all RAC determinations approving EK2-EK3 mechanisms were made available to principal investigators through the IBCs. By the time NIH was notified, the controversial experiments had ceased, but he had ordered further inquiries nonetheless.[56]

A second meeting with RAC was now scheduled, following which the director could place his proposals in the *Federal Register* for comment. The burden of proof had now shifted, Frederickson announced in his April 27, 1978, introductory remarks, to the proponents of regulation.[57] In this spirit, his technical advisors quickly approved most of the changes enhancing flexibility. As for nonscientific issues, they agreed that the IBCs could trigger alterations in research protocols without Bethesda's prior approval; after all, this was precisely the division of labor ordained by the Commission on Human Subjects. And when political scientist Emmett Redford, one of RAC's "public" representatives, suggested that each IBC be instructed to include two members from outside the university community, Professor Edward Adelberg countered that since this was not a question of science the committee should steer clear of offering opinions. RAC contented itself with advising the IBCs to consider public representation.[58]

The director's revised guidelines and his justifications for these were disseminated on July 28, 1978, and contained the following salient points. No longer would all modes of recombinant DNA research be pigeonholed into a set of admittedly rigid categories. The British "case law" approach, affording both exceptions to specific recombinations on the prohibited list and total exemptions to various other experiments (such as, for example, where the exchange of genetic properties was known to occur in nature), would now furnish RAC with a discretion hitherto lacking. Further flexibility was provided through an expansion of available cloning vehicles, as HV1 options other than *E. coli* K-12 could now be employed. And no longer, Frederickson asserted, should substantive goals weigh less heavily than procedural niceties in the ad-

56. Ibid., 549–54.
57. Ibid., 519.
58. Ibid., 528.

ministrative process.[59] Copies of all new NIH grant proposals would henceforward be routed from the IBC to the Office of Recombinant DNA Affairs (ORDA) at Bethesda for both registration and approval of guideline requirements. Hence, the study sections manned by academics could concentrate solely on the scientific merits presented. Moreover, ORDA need apply only a postreview check on twelve-month reapplications and protocol changes, unless P4 investigations were involved; the IBCs had seemingly shown themselves amply qualified to prescreen all other overtures.[60] However, biosafety panels were obligated to choose at least one member from the larger community, and the guidelines (except for the provision mandating final NIH authorization) would be extended to cover all cloning projects sponsored by institutions where *any* of its personnel received agency support for gene splicing. The former provision was almost certainly an attempt to head off the proliferation of city censorship boards,[61] while the latter amendment was defended on grounds of safety and consistency.[62] Finally, the director addressed special attention to what he called the "due process" component. By this he did not mean the right of scientists to fair procedure should their grants be threatened with revocation but, rather, the "right of the public" to participate in the formulation of guidelines. For those who believed NIH had not conducted enough open hearings and had not provided enough representation for interested and affected groups, Dr. Frederickson emphasized that all meetings of the RAC and of the Director's Advisory Committee had been accessible and that all proposed revisions had been published for outside notice and comment. He left it to his superior, HEW Secretary Joseph Califano, to observe that the NIH had held nineteen hours of public hearings on the issue of guideline amendment, and that the department's general counsel would conduct yet another such meeting just to ensure complete access.[63]

Working one's way through the minutes of the general counsel's September 15 meeting is an exercise in *déjà vu*, though not nearly so

59. Ibid., 6. The director's "Proposed Revised Guidelines" are found in the *Fed. Reg.* 43, no. 146 (July 28, 1978): 33042–33178.

60. Technically, most new proposals might also commence following IBC concurrence, but the monies would have to come from other sources because "prior to award, MUA's will be reviewed by ORDA for [guideline] compliance." *Recombinant DNA Research* 3 (September 1978): 53; *Fed. Reg.* 43 (July 28, 1978): 33092. In accord are the December 22 "Revised Guidelines."

61. *Recombinant DNA Research* 3 (September 1978): 26; *Fed. Reg.* 43 (July 28, 1978): 33065.

62. *Recombinant DNA Research* 3 (September 1978): 10; *Fed. Reg.* 43 (July 28, 1978): 33049.

63. *Recombinant DNA Research* 3 (September 1978): 3, 27; *Fed. Reg.* 43 (July 28, 1978): 33042, 33066.

tedious as having to conquer hundreds of repetitive three-column pages of guidelines in the *Federal Register*. The only new wrinkle was the growing militance of some environmental groups and their allies, evidently desperate and certainly frustrated as deregulation time neared. RAC was hopelessly unrepresentative of *vox populi*, they asserted; one-third of its membership should come from labor, public interest organizations, public health officialdom, and the OSHA, EPA, FDA bureaucracy. Peer review had also failed, as guideline violations at Harvard Medical School and the University of California at San Francisco clearly demonstrated; IBCs must be revamped so as to better reflect the interests of the people with the most to lose and the people with the most to gain from this research. Moreover, no guideline exceptions for prohibited work ought to be granted unless a full environmental impact statement had been rendered. Congress could not enact the sweeping regulations that were needed because of "hardball politics"; but that was no reason why HEW should not at once place recombinant DNA investigations under the roof of section 361 of the Public Health Service Act, treating the research for what it was, a clear danger to the spread of communicable disease.[64] Though the legal and constitutional dimensions of these arguments received no consideration at the time, that is a loophole deserving now of closure.

The National Environmental Policy Act (NEPA) mandates an in-depth impact analysis whenever federal governmental conduct might affect significantly the public's environmental interests. NEPA's stock-in-trade is technology assessment, viz., formulating a cost-benefit calculus based on estimates of risk. While NIH actually provided an environmental impact statement for its 1976 guidelines, the fit between laboratory research and environmental consequences can at best be called tenuous. Unless a clone escapes into the outside world, survives, and multiplies, there is no impact. And much recombinant DNA experimentation is not technology; it is quasi speech. Not even drive-in movie-house owners who exhibit X-rated films have ever been asked to file environmental impact statements![65] To make "times, places, and manner" funding strings contingent upon proofs of insignificant environmental harm—where the alleged danger is expressive activity carried on behind closed doors—cannot survive constitutional scrutiny. The director rejected that invitation, relying on the fact that he had submitted instead the less demanding, but more than adequate, environmental impact assessment.

In order to trigger section 361 of the Public Health Service Act, there must be a reasonable basis for finding that particular substances, orga-

64. *Recombinant DNA Research* 4 (December 1978): 107–8, 150, 171, 173, 176.
65. Cf. Erznoznik v. Jacksonville, 422 U.S. 205 (1975).

nisms, and the like are pathogenic to man. To some proponents of recombinant DNA regulation, the widespread debate over health and safety issues itself established sufficient grounds for section 361 coverage of cloning throughout the nation. But to quote molecular biologist Norman Zinder, "This may be good politics . . . but it is certainly not good science,"[66] for the record fails to show that the typical gene-splicing exercise produces a disease-spreading artifact, much less that all clones are pathogens. But the contention is also not very good constitutional law. Clearly, Congress' power to curtail hypothetical contagion provides no greater justification for saddling ways of knowing with overly broad restrictions than do other plenary authorities at the legislature's disposal. Secretary Califano vetoed the environmentalists' recommendations, citing their marginal scientific basis.

The issue of public input, however, proved to be a stickier wicket. With good reason, NIH seems to have been defensive about the Recombinant DNA Advisory Committee, which was supposed to devote its energies to arcane scientific matters, but which every now and again strayed into other pastures. Part of the problem, of course, was that cloning does not subdivide neatly into categories of expertise; biological questions can quickly become political and legal questions. Acting on that dilemma, HEW decided to expand RAC's membership to twenty-five panelists and forthrightly broadened the board's jurisdiction to relevant nonscientific concerns. Newcomers included professors of law, education, environmental studies, occupational safety, as well as a retiring member of Congress. Joining forces with two holdovers—academicians versed in political science and bioethics—this seven-person group would now "represent" interests beyond the world of hard science.

Much touchier was the question of IBC makeup. It was one thing to restructure in-house policymaking, but quite another thing to tell academic institutions who should supervise specific research projects. NIH, however, had already opened the door by proposing one outside member per biosafety panel. This, the public interest community had argued in effect, was tokenism. Very well, Director Frederickson countered, we will raise the figure to two, or a minimum of 20 percent of a committee's total membership, the same percentage assigned to RAC laypeople. And so with release of the new guidelines, Secretary Califano was able to say that the relaxation of constraints had been linked to a significant increase in public participation respecting both program formulation and compliance.[67]

66. *Recombinant DNA Research* 4 (December 1978): 166–67.

67. Ibid., 3. HEW's "Revised Guidelines" are in *Fed. Reg.* 43, no. 247 (December 22, 1978): 60080–60130.

The basic political question is, Why should Harvard and Berkeley feel any sense of responsibility to local attitudes, where these are contrary to the schools' conception of scholarly obligation? The basic constitutional question is, Assuming that the greater good requires such responsiveness, by what means can the state best achieve harmony through law between scientific inquiry and communal concern? As to the first point, academic institutions are not islands, and they well realize that they must either honor applicable statutes and ordinances or make many enemies by going to court and getting them overturned. But to argue from this that they owe surrounding communities representational access to their decisionmaking bodies, especially when the subjects for discussion may well involve the guts of academic freedom, can hardly be accepted at face value. It is indeed curious that the revised guidelines provided researchers with no right of appeal should the petitions they place before their IBCs fail of approval; after all, said Director Frederickson, NIH does not wish to encourage adversarial relations between investigator and institution.[68] But an institutional biosafety committee, with a 20 percent "outsider" membership quota mandated via federal order, has ceased to be merely the institution. It has become the institution plus. And this plus factor turns committee deliberations into an adversarial context by definition. All of which serves to raise the second point. If American national government, as granting agent, has the constitutional power to formulate the NIH guidelines, then national governmental officials can take the next step and place *their* people on the committees which oversee those guidelines. The price, however, may be a rather high one. When a university obeys state law enacted pursuant to the Constitution, it upholds larger principles of republicanism. When the government infiltrates the university by "purchasing" decisionmaking prerogatives, it threatens to erode the free expression networks that republicanism is designed to protect.

On December 22, 1978, the NIH's Revised Guidelines were at long last promulgated. Containing all of the provisions floated out for comment four months earlier as well as the "participatory" guarantees just described, they had the effect of exempting from administrative rules and controls approximately one-third of NIH-funded cloning investigations. They also reduced by at least one level the physical and biological containment criteria imposed in 1976 as these had now been reevaluated and reapplied.[69] After an examination of the published data and opinions, it is difficult to label those deregulations anything but salutary. Almost as

68. *Recombinant DNA Research* 4 (December 1978): 17; *Fed. Reg.* 43 (December 22, 1978): 60094.
69. *Recombinant DNA Research* 4 (December 1978): 3; *Fed. Reg.* 43 (December 22, 1978): 60080.

an afterthought, the director noted that the great legal, ethical, and social issues raised by recombinant DNA research had gone largely unmentioned in his document. It would appear, he offered, that these concerns warranted study.[70]

Even suspicious environmentalists might well have expected 1979 to be a quiet year at NIH headquarters, while the new guidelines were given a chance to sink in. But as early as April, a notice appeared in the *Federal Register* soliciting commentary on a proposed exemption for all experiments utilizing EK1 and EK2 host vectors. This proposition a RAC study group watered down somewhat; still, the suggested revision embraced the vast majority of known *E. coli* K-12 manipulations. By the time the full panel convened in September, fourteen communications of strong support for drastic change lay on the table, including one signed by 183 specialists. Risk-assessment evidence was presented there, showing, among other things, that when polyoma virus was cloned in a pBR322 EK2 plasmid and the hybrid was then injected into mice, infection failed to develop. Suppose, however, an EK1 host had been employed, a panelist inquired. No cause for concern, was more or less the reply: plasmid transfer did not materialize in any of the tests, and besides, the really suspect exercises, such as recombining known toxins, would continue to be prohibited. How much research did the minigroup recommendation encompass, asked another committee member. Approximately 85 percent of investigations then in progress was the answer. But of course, observed another supporter, the proposal was bound to save cloners lots of time! Finally, by a vote of 10 to 4, RAC approved the motion as amended. Put concisely, its action freed most *E. coli* K-12 recombinant DNA work from the main thrust of guideline coverage, prescribing only P1 restrictions for all such endeavors and requiring principal investigators to file with their IBCs only registration documents spelling out research particulars. However, campus officials would have no authority to exercise prior restraint over these experiments.[71]

This and other authorized exemptions and exceptions were not without their repercussions within the RAC family. It seemed, a member reflected, that the committee's focus had undergone unannounced change. Heretofore, cloning was regarded as a species apart and since some dangers obviously could not be anticipated, preventive measures seemed desirable. Now, the guiding principle had become scientific expediency. If someone wanted to do something and specific dangers could not be documented, then the group was allowing research to go forward. Our

70. *Recombinant DNA Research* 4 (December 1978): 26; *Fed. Reg.* 43 (December 22, 1978): 60103.
71. *Recombinant DNA Research* 5 (March 1980): 154–65.

rules have become shot through with loopholes, some lamented. We need *new* rules, others replied. In fact, gene-splicing innovations were developing at such a rapid pace that the December 1978 restrictions had already begun to squeak at the hinges; yet, RAC was not really prepared to face the consequences, nor did it even possess the political tools to respond effectively to them. When a motion was introduced that would have altered officially the committee's overall standard of review to include only estimates of meaningful, realistic hazard, panel members balked. How can we posit a general criterion of perceived hazard, they asked, and then proceed to enumerate a series of guidelines predicated on unperceived hazards? RAC was clearly impaled on the horns of a blueprint strategy which others stood ready to defend.[72]

The proposed changes for *E. coli* K-12 investigations, moreover, ran into stiff opposition, as these larger political forces rushed in to defend the breach. Roy Curtiss, who had labored for so long to develop the ultrasafe chi-1776 EK2 vehicle, wrote a detailed rebuttal, asking essentially why EK1s were suddenly so innocuous. On the frenetic wing, *New Scientist* magazine excoriated RAC's actions because only ten of twenty-five members had signaled their approval, not even 50 percent of the total group. And why, Senator Adlai Stevenson asked newly appointed HEW Secretary Patricia Harris, was NIH giving up so soon on its new guidelines?[73]

To Director Frederickson, the RAC recommendation seemed, in the light of all this, a tad precipitous. Hence, he determined that the *E. coli* K-12 research at issue should be classified P1-EK1, and in the special case of experiments where the purpose was genetic expression of a eukaryotic protein, he accorded to IBCs the power of prior approval with an option to add restrictions as they saw fit. Undoubtedly the cloning community would have lived with this modest reassessment, but when Secretary Harris insisted to the director that another thirty-day comment period was in order, even though the deadline had passed under revised NIH rules, those who had led the charge at Asilomar were beside themselves. "Public process" had now become a millstone around the necks of scientists, wrote RAC member and Nobel Laureate David Baltimore. The RAC vote, had it been held in the United States Senate, would have been sufficient to ratify a nuclear disarmament treaty, argued an exasperated Norman Zinder. And Stanford's Alan Campbell, who earlier had opposed a change in RAC's overall "watchdog" approach, was now prepared to rewrite the guidelines from stem to stern, making them applicable only when "hazard is clearly perceived." All in all, 169 out of the

72. Ibid., 180–81, 410.
73. Ibid., 339–40, 364–66, 374.

185 letters conveyed to Donald Frederickson on this subject supported his findings, which went into effect on January 29, 1980.[74] If Patricia Harris had wanted scientific analysis by plebiscite,[75] she, in a sense, had achieved it. But, the NIH director noted casually, this unprecedented call for public assistance would probably never be repeated if he had anything to say about it.[76]

By no means did these considerable changes pull the teeth from NIH oversight, as subsequent events would show. In fact, though, Bethesda has found, following careful investigation, only four instances of rule-breaking. One, involving a UCSF team and the improper deployment of a particular vector plasmid, we have already mentioned. Earlier, in December of 1977, NIH suspended Charles Thomas' research at the Harvard Medical School because proper documentation for his experiments had not been filed. Both these matters had entailed merely "administrative violations," according to Director Frederickson.[77] Causing larger waves was an incident occurring in 1980, where the principal figure was Ian Kennedy of the University of California at San Diego. He had utilized a prohibited virus, had had his cloning endeavors suspended by the campus biosafety committee as soon as the panel had heard a report on his actions, and had subsequently resigned his post at the university.

Far more controversial, and certainly more complicated, was the behavior of UCLA's Martin Cline. He will go down in medical history as the first person to attempt to transplant cloned genes into human patients, his particular purpose being to effect cures for the dreaded globin disease beta thalassemia. These experiments he performed in Italy and in Israel, but his initiatives ignored salient controls established both here and abroad. Stated briefly, Cline failed to obtain approval from either the UCLA biosafety committee or the UCLA human subjects review board to perform such experiments overseas; he received no green light from Italian human subjects officials; and he did not inform Israeli personnel that his therapy involved gene splicing. He did, however, receive informed permissions from the patients themselves and their families. Nonetheless, NIH investigators did not respond mercifully, meting out three noteworthy sanctions. First, Cline could do no further experiments with either human subjects or nonexempt DNA recombinants unless Bethesda approved these in advance. Second, should he apply for more research support, his disobedience would be considered part of his *curriculum vitae,* an onerous burden to overcome indeed. Third, a review

74. Ibid., 11–14, 381–82, 435, 451–52.
75. Ibid., 502–3.
76. Ibid., 13.
77. *Fed. Reg.* 43 (December 22, 1978): 60092.

panel would examine his four ongoing grants and decide if they ought to be continued. Six months later, NIH announced that Cline had been stripped of funding for two of these projects. While some responsible critics feel that this is a classic case of a researcher going astray in a bureaucratic maze,[78] NIH officials could rightly point to the fact that Dr. Cline was dealing with humans, not bacteria; that he was essentially prescribing treatment, not doing basic science per se; and that his behavior might have had serious foreign policy implications. If Bethesda intended the episode to be an object lesson, it could point to the fact that no further guideline violations have hit the headlines since.

The year 1979 saw the recombinant DNA debate take on an added dimension, as RAC proceedings clearly manifest. A number of private firms, realizing that immense profits awaited those inventive enough to convert clones into machines for mass-producing scarce resources, were now attracting national attention as the investment magnates of the future. Some of these companies, for instance Eli Lilly and Du Pont, were familiar names. Others, such as Genentech, founded in 1976 by venture capital professional Robert Swanson and biologist Herbert Boyer, were specializing in the new technology. Because all these businesses funded their own research, the NIH guidelines had no legal impact on what they did and how they did it. Needless to iterate, several public interest factions found this autonomy entirely unjustified and were quick to tell Congress about their grievances. But as has been seen, the legislative process had failed to grapple effectively with cloning as science; and that process now showed itself incapable of grappling even with cloning for dollars. If national action were to be forthcoming, which certainly seemed a sensible course, then NIH, many believed, would have to do something.

Actually, industry had powerful incentives for working up a *modus vivendi* with Bethesda. For one thing, an alliance with NIH might very well head off federal statutory regulations, should Congress ever decide to confront the subject. Secondly, ties with NIH meant access to the academic world, home of the nation's most reputed gene splicers. So the question became one of how to grant "diplomatic recognition."[79]

When RAC first addressed this issue in May of 1979, it simply passed the buck by voting—9 yes, 6 no, 6 abstaining—to commend mandatory compliance, though who would make that wish a reality remained

78. Nicholas Wade, "Gene Therapy Caught in More Entanglements," *Science* 212 (April 3, 1981): 24–25; Wade, "Gene Therapy Pioneer Draws Mikadoesque Rap," *Science* 212 (June 12, 1981): 1253; Marjorie Sun, "Cline Loses Two NIH Grants," *Science* 214 (December 11, 1981): 1220.

79. *Recombinant DNA Research* 3 (September 1978): 222.

unclear. (It would be an interesting and helpful exercise to analyze this and other cleavages of opinion among RAC membership, but the minutes do not carry these data unless individual panelists ask the record to indicate their preferences. Hence, concerned citizens often cannot know whom to praise or blame, and social scientists are deprived of the chance to apply their statistical tests of probability to RAC voting patterns. Such are some of the infirmities of the new "procedural due process.") The fact was, however, that the committee had already opened the door. Under the 1978 guidelines, experiments involving more than 10 liters per vessel were prohibited, except that the director could grant exceptions where the clones were "rigorously characterized" (that is, their base pairs had been sequenced) and were free from harmful genes. RAC was to test that criterion against each request for large-scale investigations and establish physical containment standards for these. But it was business interests, not academic interests, that had a compelling need to employ mass-quantity modes of production. So RAC had well involved itself in evaluating applied cloning. Incidentally, the 10-liter ceiling was waived with respect to *exempt* investigations, another reason the director had placed P1-EK1 constraints on the typical *E. coli* K-12 endeavor, as described above.

Convinced that voluntary compliance on the part of industry was the only viable option, Dr. Frederickson developed a draft supplement to the guidelines, the central features of which were the following: (1) participating firms must maintain IBCs, vested with the same authority and meeting the same outside membership criteria as campus committees; (2) these firms could register their projects with the Office of Recombinant DNA Affairs (ORDA) after they had cleared IBC review; (3) companies could also apply to RAC for guideline exceptions as university researchers were doing; (4) NIH would guarantee the privacy of trade secrets in examining industrial requests. By the surprisingly lopsided tally of 11 to 0, RAC agreed to these stipulations. At its September 1979 meeting, the committee for the first time went into closed session, approving a submission by Eli Lilly to clone large-scale amounts of *E. coli* for insulin production, on the condition that NIH be accorded the privilege of site inspection.

Consensus began to come apart in December, however, when RAC was apprised of a joint recombinant DNA transaction involving the USDA and Genentech. The former maintains jurisdiction over all studies utilizing the highly pathogenic foot and mouth disease virus and these experiments are restricted to Plum Island, New York. Genentech wanted to develop a vaccine for the sickness. As this virus has an RNA-constituted genome, the plan was to subject that genome to reverse transcriptase, i.e., to convert artificially the RNA to DNA, clone various of the DNA fragments in *E. coli* K-12, and ship them for sequencing to Genentech's laboratories in San Francisco. Since splicing the genes of lethal animal pathogens was forbidden, the petitioners were requesting that an exception to the guidelines be permitted. Why must the artifacts be sent cross-country, Harvard's Richard Goldstein asked. Because the USDA does not maintain a recombinant DNA facility at Plum Island, came the reply. Can't vaccines be produced by means other than cloning, Tufts' Sheldon Krimsky inquired. Tissue culture systems, he was told, are much too expensive. The key vote turned on whether containment at Genentech should be P1-EK1 or either P3-EK2 or P2-EK2. The less stringent option prevailed, with 9 yes, 7 no, 2 abstaining.[80] But again Director Frederickson temporized, explicitly reserving judgment on the San Francisco research until the Plum Island aspects had been completed.

At the December 1979 RAC assembly, Professor Krimsky launched a broadside against the notion of voluntary compliance. How do we know whether a company has registered any or all of its experiments? Suppose a firm cheats, how do we "punish" the violation? To this, industry representatives stressed that OSHA not only has authority to regulate the workplace but inspects their facilities constantly. As a matter of fact, OSHA generally confines agency investigations to "recognized hazards," and research with recombinant DNA, as its advocates have so often contended, poses only theoretical or hypothetical dangers.[81] That was why OSHA's own leadership had supported the call for NIH voluntary scrutiny. Another argument put forward by company representatives was that cloners would be deterred from cheating because of the status risks involved. Yet we have already seen how the impetus to publish has caused its share of data falsification. For the moment, voluntarism ruled the day only because everyone knew Congress couldn't or wouldn't act and because the program was specifically labeled experimental and subject to further review.[82]

During 1980, RAC's role as orchestrator of industrial scale-up became a dominant point of disputation, with the larger question of NIH jurisdiction over commercial cloning providing the inescapable backdrop. By this time, the panel had endorsed a new set of physical containment standards—i.e., P1-LS, P2-LS, P3-LS (LS = "large scale")—for experiments employing more than 10 liters. To take care that plant equipment met these standards, RAC was now exercising prior review of

80. Ibid., 5 (March 1980): 397–405.
81. Ibid., 171–72.
82. Ibid., 416–20.

specific proposals and had even participated in a site inspection of the Eli Lilly facility. But in the opinion of the Coalition for Responsible Genetic Research, which numbered among its members George Wald, Jonathan King, Francine Simring, Liebe Cavalieri, and Sheldon Krimsky, large-scale investigations without any risk assessment whatever to measure their special consequences could well result in another Three Mile Island. And an AFL-CIO representative also offered criticism, contending that RAC's job was research oversight not workplace regulation, and that labor had no voice whatever in RAC decisionmaking.[83] However, Krimsky's motion to defer to OSHA on large-scale industrial cloning activities was tabled, 8 to 6.[84]

Correctly sensing committee disarray, Director Frederickson made an appearance at the June meeting, where he reiterated his favorite theme of a single national agency formulating guidelines rather than regulations. Let's not delegate our scientific know-how to anyone else, he was saying; and by all means, let's not give the politicians an excuse to resurrect their statutory panaceas. If panel members feel they are in over their heads, then RAC should stick to biological and P-LS assessments and leave machinery oversight to others, such as the company IBC. It is no sin to examine confidential industrial information, he concluded, if we don't stray from our own area of expertise.[85]

To Susan Gottesman, an NIH scientist and RAC member, these suggestions hit the mark and she moved their adoption. To David Baltimore, RAC should stand by its guns and not convey for a moment the message that industry was being permitted to do its own thing; his fear was political reprisal that would damage the new science and he obviously considered the question of RAC's "unclean hands" insignificant. To Sheldon Krimsky and Richard Goldstein, who had earlier in the discussions disparaged RAC's attempts to do what they thought OSHA should be doing, it would be most unfortunate for the committee to retreat from the concept and responsibility of prior technological review. But Dr. Gottesman won the first round, when her proposal achieved success by a 12 to 5 count. That is, RAC would now determine whether large-scale numbers of recombinants were sequenced and benign, specify the applicable P-LS experimental level, and rely on the local IBC for laboratory monitoring. She captured the second round even more decisively, for when it became clear that a majority favored establishing an oversight subcommittee to advise the NIH director on standards and facilities relating to the mass production of cloning, a motion by Krim-

83. Ibid., 6 (April 1981): 259–60, 289, 433–36.
84. Ibid., 67.
85. Ibid., 104–6.

sky to reintroduce the concept of prior review was trounced, 15 to 3. Though some firms found this last supervisory thrust a form of discrimination because it presumed large-scale endeavors to be riskier than smaller ones, Dr. Frederickson had no difficulty accepting the entire package.[86]

Tracing these hours-long debates, keeping careful track of which amendments were being offered to what motions, the reader might well be inspired to ask why these intellectuals, most of whom were distinguished scientists, were expending so much energy on a policy issue which had practically nothing to do with the nature and consequences of scholarly inquiry. We live in an era when Congress regulates the industrial sector almost as a matter of course. Why could not the legislative branch, without posing even an imaginary threat to academic interests and NIH hegemony over those interests, have enacted appropriate measures setting out—or instructing the administrative establishment to set out—reasonable physical containment criteria for the large-scale commercial production of clones?

Part of the problem, as usual, was Congress' own lack of focus. On this occasion, interested members were not even considering meaningful recommendations of suspect constitutionality, such as, for example, a bill which would treat all recombinant DNA research carried on at Genentech as though it were perforce technology/conduct, and which, therefore, lay across-the-board regulatory criteria on these investigations without even providing a rebuttable burden of proof to the contrary. Instead, the indefatigable Senator Stevenson, who had now seemingly replaced Senator Kennedy as unofficial upper chamber watchdog for DNA experimentation, was promoting a notification statute that would mandate all cloners not employed by an institution receiving NIH money for gene splicing to file with HEW Secretary Harris complete data on their investigations.[87] Just what Ms. Harris was supposed to do with this information remained uncertain, though the threat of sanctions might flush out some researchers who had failed to register with ORDA. One larger consideration governing congressional inertia seems to have been that the lawmakers saw themselves as Bethesda's handmaiden with respect to recombinant DNA politics, even though the NIH and Genentech lived in two different constitutional worlds, neither owing the other a measure of reciprocity or obligation. Perhaps an even more important factor making for legislative quiescence was testimony previously noted that federal inspectors would have difficulty searching out undefined biohazards. Of course, Congress could specify standards for equipment

86. Ibid., 108–12, 135–37, 164–66, 175–79, 320–22.
87. S. 2234, 96th Cong., 2nd sess. (1980).

design without having to wait for a clone to escape. But what the politicians were beginning to see was that, from the safety angle, the recombinant DNA debate was looking more and more like a tempest in a teapot. Instead of adding another niche to the bureaucratic state, maybe the thing they ought to do was let the whole business fade away.

Almost lost in the protracted attention accorded industrial cloning machinations was RAC's overhaul of NIH's registration requirements for recombinant DNA grant-holders. As reported in Chapter 1, Bethesda, unlike NSF, has never insisted that research exempt under the guidelines be reported to either national or campus oversight bodies. The Goldstein-Krimsky bloc had moved, in December of 1979, to provide for IBC notification, but had been outvoted, 10 to 5. Opposition spokesmen made the point that if Bethesda need not be informed about P1-EK1 *E. coli* K-12 experiments, then why should local people have to know about exempt experiments? More paperwork, they felt, could be justified only on grounds of safety. During that discussion, panelists conjectured on how much cloning was then exempt; the correct figure, they thought, was between 50 and 60 percent. Note that those estimates were much lower than the 85 percent bandied about earlier, evidently because of expanding research in the eukaryotic field. The consensus simply did not consider it important that government officials, as part of their supervisory rather than their risk-protection function, even know precisely how much gene splicing they were *not* responsible for.[88]

Flushed with this success, the Asilomar establishment now went after what it had come to regard as the twin bureaucratic horrors of prior review and central registration of experiments. Maxine Singer took the lead, petitioning RAC to wipe the term "memorandum of understanding" from the guidelines and to put upon principal investigators the sole obligation of telling their IBCs what they were doing. The appropriate model should now be national constraints implemented by local bureaus, the argument ran, for federal project review meant little without federal surveillance machinery. The demise of ORDA's censorship authority was easily accomplished, but a middle-of-the-road group joined Goldstein, Krimsky, et al. to maintain prior consultation with institutional biosafety committees. The issue here was no great dispute over the principle of previous restraint; it was, rather, a question of perceived IBC competence. To the Singer forces, a considerable burden of proof lay with the IBC vis-à-vis the principal investigator, while, to the public interest forces, a considerable burden of proof lay with the IBC vis-à-vis Bethesda. But the "center of gravity" proponents submitted that local authorities had shown themselves knowledgeable and responsible enough

88. *Recombinant DNA Research* 5 (March 1980): 165–66, 413–14.

to be free from Bethesda's nipple and argued that they were certainly better suited to make informed, objective decisions than researchers themselves. In fact, there was no persuasive evidence in the record regarding the several aspects of IBC performance, as RAC itself would acknowledge shortly thereafter. The panel then concluded its deliberations by repealing the requirement for central registration. So while national guidelines continued to obtain, national headquarters would henceforward keep no official records of who was constrained by them.

Seventeen letters, sixteen of them favorable, arrived at NIH during the public comment period. Quite understandably, Paul Berg found it absurd that scientists had been forced to file revised memoranda of understanding and agreement every twelve months with both their IBCs and ORDA, and he was glad to see that process curtailed. Norman Zinder felt that at long last the guidelines would now be guidelines, not regulations. The lone dissenter was the American Society for Microbiology, which argued that central registration was defensible simply because some federal official ought to know who was being covered and why.[89] Indeed, if we put the issue of hazard to one side, how could any meaningful policy assessment of the guidelines be developed, how would any legitimate study of recombinant DNA research as a social, as a scientific, or as a *political* process be feasible if federal officials kept no files on experiments they considered worthy of unique legal oversight? In that regard, one's attention is drawn to a comment offered by Johns Hopkins Nobelist Daniel Nathans, who said, "At Johns Hopkins University the [IBC] functions extraordinarily well, and its role in the review of proposals is highly respected."[90] Those who are concerned with studying the social science–natural science recombinant DNA mixture will be pleased to hear that news. But I am constrained to report that when I endeavored to interview members of the Johns Hopkins community on precisely these matters, I was denied entrée.[91] Harsh empirical realities, of course, can also come at the hands of Washingtonians; however, it is somewhat easier to obtain accountability from one office than from hundreds. But no doubt those considerations were far from Dr. Frederickson's mind when he accepted all these RAC initiatives.

With the onset of the 1980s, the focus of recombinant DNA disputation has shifted somewhat from safety issues to money issues, from federal guidelines orchestrating the ground rules of funded research to

89. The preceding discussion is based on *Recombinant DNA Research* 5 (March 1980): 231, and 6 (April 1981): 168–70, 184, 210–12, 291, 303, 314.

90. Ibid., 6 (April 1981): 296.

91. Appendix A, below, describes the sampling procedure upon which the interviews presented in Chapter 4 are based.

campus-corporate agreements that, in some cases, threaten to alter drastically traditional academic workways. The laboratory breakthroughs spawning these provocative arrangements came as early as 1977, when a team of California scientists successfully instructed *E. coli* to produce the hormone somatostatin. Building on experimentation at Harvard, which had cloned the gene for rat insulin, this group then contrived bacterial hybrids capable of manufacturing human insulin. The biotechnological announcement of 1980 was that researchers had synthesized human interferon, an important antiviral protein. Since then, the tools of microbial expression have yielded potential vaccines for foot and mouth disease and hepatitis B. As the list of genes which have now been linked to specific areas of man's chromosomal chart grows monthly, so also does the list of genes expressed through recombinations in both prokaryote and eukaryote hosts require constant updating.

The work involving insulin and interferon is especially instructive. First of all, both are exceedingly precious, hence expensive, disease fighters. If one could mass-produce either for human consumption, the profits would be enormous. Second, significant aspects of this research have important implications for basic science. Recombinant DNA investigations have shown that man produces not merely a few varieties of interferon, but, rather, more than a dozen subtypes. It was a signal triumph when cloners located the gene for the unusually potent and rare gamma interferon in human white blood cells and managed to induce its manufacture. Moreover, both human insulin and human interferon genes have been transplanted into fertilized mouse egg cells, providing important steps in our understanding of genetic control systems. Finally, many of these experiments—including several exercises in what can only be classified as pure research—were accomplished either in the laboratories of commercial outfits or by teams some of whose members had strong financial ties with profit-oriented firms. For example: future Nobelist Walter Gilbert, who headed up Harvard's rat insulin project, had already become a founding father in the Swiss genetic engineering company Biogen S.A.; human insulin cloning was accomplished through a collaborative effort involving scientists at City of Hope Medical Center and Genentech; and the gamma interferon research was carried out in-house by Genentechers alone.

Given this background, a long-term pooling of academic and corporate resources seemed, to many participants, eminently reasonable and rewarding for all concerned. Universities had the brainpower and the prestige, but they were running behind in funding; business interests had the capital and investment know-how, but they needed access to the latest scientific advances. Colleges might pride themselves on being forums for intellectual discourse, yet this had never stopped them from building their endowments through the medium of shrewd investment decisions. Faculty might lust after professional rewards, but this had never precluded them from increasing their salaries by doing outside consulting for industry. And both school and teacher had for years been known to collect royalties from patents obtained on inventions honed in the campus laboratory. Private enterprises, on the other hand, had many times been willing to stake academic research in the past, and they often were rewarded with first options on patent licenses when educators found it expedient, as they sometimes did, to negotiate these quid pro quos. As for the public, it stood a good chance to reap cloning harvests in new drugs, cheaper, more abundant foodstuffs, and genetic therapy techniques.

However, the first well-publicized joint cloning venture boomeranged.[92] Harvard, which could never have expected to make much money out of Walter Gilbert's accomplishments, given his Biogen connection, proposed to guard against repeat episodes by purchasing a minority equity position in a new gene-splicing firm. Other stockholders were to include venture capitalists, who would supply funds; top management, which would run the show; and recombinant DNA researchers, who would, as a rule, work for the company, but at least one of whom, Mark Ptashne, was a star Harvard molecular biologist, possessed of cloning expertise sufficient to portend healthy profits. Harvard would own all patentable artifacts arising from Professor Ptashne's investigations but would issue licenses for their development to the firm itself. Nothing in the agreement limited Ptashne's freedom to publish. Why did the plan fail to get off the ground? Of all the explanations advanced, the key phrase seems to have been "conflict of interest." Harvard's faculty was unable to accept an arrangement whereby one of its own would hold hands with the university proper in a grand enterprise not to extend the bounds of knowledge but to enhance their respective pocketbooks.

It would be impossible to enumerate the central features of the typical recombinant DNA research arrangement between the worlds of commerce and academia, in part because a complete list has never been compiled and also because a blanket of confidentiality has been thrown around many of the contracts known to exist. This chapter concludes with a discussion of three of the most important such agreements, em-

92. For the facts surrounding the Harvard fiasco, see Michael Vermeulen, "Harvard Passes the Buck: The DNA Affair," *TWA Ambassador,* January 1982, 41–57; Barbara J. Culliton, "Biomedical Research Enters the Marketplace," *New England Journal of Medicine* 304 (May 14, 1981): 1195–1201.

phasizing their political and legal implications and especially addressing their potential impact on the campus as a center for competing ideas.

In September of 1981, Stanford and the University of California at Berkeley entered into a unique funding transaction with a new biotechnological company, Engenics.[93] The firm, coestablished by chemical engineers holding professorships at the two institutions and a group of businessmen, had raised $10 million from six major corporations, among them Bendix and General Foods. Under the agreement, Engenics pledged 30 percent of its stock to a nonprofit center, which would disburse research support to participating academic units. For starters, the center was giving the two California schools $2.4 million for cloning research. All commercial development activities stemming from campus endeavors would be the firm's responsibility and source of income. Thus, participating companies had the unprecedented opportunity to acquire considerable equity at ground floor prices in university research projects, while the campuses not only received a nice gift but also a chance to share in further bounties should the outfit's profit margin so warrant.

The Engenics model is a deliberate attempt to allot home institutions a piece of the recombinant DNA pie and at the same time avoid the pitfalls of the Harvard joint-shareholder scheme. Were campus faculty personnel uninvolved in company directorship, a presumption of inadequacy might seem ungenerous indeed. But over and above any financial benefits accruing to Stanford and Berkeley, there is no escaping the fact that professors—aided and abetted by university fathers—are here investing in the research of *some* of their colleagues, risking their capital in the expectation that financial payoff lies at the end of particular scientific rainbows. For it is eminently clear that many of the most valuable long-range scientific investigations—inquiries which lie at the core of First Amendment concern—can never produce commercially feasible spin-offs, can never serve as a source of financial return to the expectant entrepreneur. Whether one believes the university environment can accommodate this sort of arrangement or not—a normative question we face in Chapter 5 —these contractual terms hardly dispense the root conflict-of-interest questions that troubled Harvard's faculty.

Four months earlier, the press had trumpeted news of a seemingly stupendous industrial subsidy for gene-splicing experimentation.[94] Hoechst A.G., a West German drug and chemical company with 486

93. Ann Crittenden, "Universities' Accord Called Research Aid," *New York Times,* September 12, 1981, 32.

94. Jane E. Brody, "U.S. Hospital Gets $50 Million Grant from Abroad," *New York Times,* May 22, 1981, A1, B15.

subsidiaries in 131 nations, had agreed to allocate $70 million[95] over a ten-year period to the Massachusetts General Hospital, to that time the largest research grant ever bestowed upon a nonprofit scientific institution in the United States. To be sure, MGH is not a degree-granting college or university, but it is a teaching and research arm of the Harvard Medical School. And the money would be used to launch a department of molecular biology under the direction of Howard Goodman, who was being wooed away from the University of California at San Francisco. That Goodman had every intention of building a first-rate center for basic scientific studies was evident. In fact, the agreement specified that MGH geneticists could select their own research subject matters, collaborate with outsiders as they saw fit, and maintain freedom of publication. Hoechst would receive in return the right to send personnel to the new shop for training, exclusive licenses for patent exploitation should appropriate inventions or discoveries be made, and access to the latest theoretical findings and innovations. MGH, as prospective patent-holder, was guaranteed royalties, though at rates that reflected the licensor's initial investment.

The Hoechst-MGH transaction poses its own set of serious interest conflicts. At that time, the hospital was the recipient of approximately $20 million in NIH support, while subsidies from non-Hoechst sources already slated for expenditure in the molecular biology laboratory amounted to $500,000. These facts did not escape the suspicious eye of Congressman Albert Gore, Jr. (D-Tenn.), chairman of the subcommittee on investigation and oversight of the House Committee on Science and Technology. To what extent, Representative Gore inquired, would the American taxpayer be funding research in an American facility only to see the know-how and the profits float off to West Germany? As it was, federal law required patent owners to give U.S. firms a competitive edge in the licensing process, where federal monies were involved. To slam the door on this line of criticism, Goodman announced that all previously solicited or earmarked departmental support would be surrendered; Hoechst had pledged to pick up the tab for shop expenses "right down to the last test tube." Even the comptroller general, in his report to Representative Gore, agreed that Goodman had solved his problems with respect to Congress' patent policy.[96]

95. At first, the figure was thought to be $50 million. For the latest word on the details of this transaction, see Barbara J. Culliton, "The Hoechst Department at Mass General," *Science* 216 (June 11, 1982): 1200–1203.

96. Richard A. Knox, "US Eyes MGH Link to West German Firm," *Boston Globe,* October 24, 1981, 13–14; Culliton, "The Hoechst Department," 1202.

But the more subtle, more far-reaching problems remain. What are we to make, for example, of the role of Harvard University in these negotiations? Again and again, the media have labeled the transaction as involving only a hospital, a corporation, and a scientist; Harvard was no party to the accommodation, university president Derek Bok has said. But Howard Goodman received a tenured appointment in the Medical School as part of the package, where he will be working in close contact with his superior, the genetics department chairman Philip Leder—who had himself entered into a $6 million research funding arrangement with Du Pont. Goodman also fully expects that his appointees at MGH will receive Harvard positions as well. How can it be that Medical School officials did not approve in advance these quid pro quos? How can it be that Harvard is not accountable for those members of its faculty who make outside bargains to spend most of their time working for an enterprise supported by a commercial outfit? If Massachusetts General's department is to be a veritable island of Hoechst-supported scientists, then what happens when Harvard-Du Pont's Leder swaps ideas with Harvard-MGH-Hoechst's Goodman? When will the trade secret Iron Curtain descend over the dialogue? How will competing nonexclusive licenses be allocated? And if MGH cloners start exchanging information with NSF-funded Harvard cloners, then what happens to that tight little island? Most serious of all, how can Goodman's operation maintain insular status and still call itself a center of learning and scholarship?[97]

The third arrangement ripe for analysis is surely the most provocative and most important to surface thus far.[98] Its architect was Edwin C. Whitehead, one-time owner of a precision instruments firm which had been sold in 1980 to Revlon at a price of $400 million. That transaction had made Whitehead Revlon's largest shareholder and had allowed him to found Whitehead Associates, a venture capital business specializing in the bioengineering field. Expanding his horizons yet further, Whitehead now struck an elaborate deal with MIT administrators which ran as follows: he would lay out $20 million to establish the Whitehead Institute for Biomedical Research in Cambridge; he would also appropriate $5 million per annum for the new institute's operating costs; when he died, his estate would tender $100 million in endowment for the facility; the institute was to be managed by a fourteen-person board, comprising a director, who would always be an MIT professor, three members chosen

97. Culliton, "The Hoechst Department," 1201–1202.

98. For commentary, see Colin Norman, "MIT Agonizes over Links with Research Unit," *Science* 214 (October 23, 1981): 416–17.

by MIT, three others selected jointly by MIT and the board itself, and seven "Whiteheaders," three of whom were the founder's children; the institute would hire and fund twenty senior scholars, who would also be accorded appointments in MIT's biology department at no cost to the university; finally, to assist the school in meeting incidental expenses arising from the construction of MIT-Whitehead bridges, a check of $7.5 million would be forwarded immediately for departmental use.

The contract, which was feverishly debated in an MIT faculty meeting but ultimately approved late in 1981 by both the professoriate and the university corporation, bristles with controversy. To Whitehead, these gifts were unadulterated philanthropy, and he emphasized that "his" institute would be a nonprofit, basic research laboratory. But just whose institute was it? If Whitehead was the only donor and his children sat on the board of trustees, why wouldn't he have access to research experimentation that both Revlon and his own company would find illuminating? But perhaps MIT had acquired sovereignty along with its financial windfall, since David Baltimore, the new director and distinguished university professor, would set the institute's research agenda and hire institute faculty. And yet Baltimore would be answerable to the board of trustees, as would his successors. So part of the ownership quandary lies in the makeup of the board, which might well split 7 to 7 on any policy matter bearing upon competing Whitehead-MIT interests. Indeed, the Founding Fathers rejected the concept of a plural executive because they feared just those kinds of logjams. In truth, the agreement, rather like the New Jersey Plan of which the two-headed presidency was a part, failed to resolve the basic political question of sovereignty, and the board's representational scheme bears witness to the abiding dilemma.

Then there is the matter of faculty recruitment and control. Baltimore argued that twenty additions to a department that would then number sixty hardly constituted a putsch, and he pointed out as well that MIT's rigorous tenure standards would have to be satisfied in arriving at all personnel decisions. But the case he presents is a bit glib. In a divided department—and MIT's biology department was very much of several minds regarding the nature of cutting edge research—increasing membership by one-third might very well prove decisive. Moreover, MIT biology would now be in the hands of two sets of caretakers, the extraordinarily well-subsidized Whiteheaders and the mere MIT'ers, who would have to fend for themselves in the grantsmanship rat race. Of course, there is nothing unusual about a faculty some of whose members have more lucrative affiliations and support structures than others. The problem is that Whitehead himself admitted he was spawning a caste system. In response to the suggestion that he simply give MIT the money and let the university proceed as it wished, he said that in *academic* institutions teaching came first, but that he wanted the Whitehead Institute to put research first. It is one thing for members of an educational unit to hold divergent views on the nature of the academic process; it is quite another thing for members of that unit to disagree on the question of whether they should tailor their primary intellectual concerns to accommodate agreed-upon academic pursuits.

Whether any educational institution can successfully mount a program of study wherein faculty missions are essentially irreconcilable depends greatly on the self-definition of that institution. What are MIT's academic purposes? How does one characterize those purposes, employing the criteria of the free expression marketplace? A capsule rendition of MIT's place in the political and economic scheme of post-World War II American life will demonstrate that the institution displays a pattern of symbiotic practice, consistently making its research services available to both government and industry for a price under contractual terms which seriously compromise its putative content neutral standing. As Dorothy Nelkin has reported,[99] the prime objective of student demonstrations in the late 1960s was to force the school to reconsider its long-standing ties with the Instrumentation Laboratory, the developer of Poseidon and MIRV hardware. Had the Vietnam War not ensued, one can seriously ask whether divestiture would have occurred, then or now. However, MIT continues to maintain close relations with the Lincoln Laboratories, in part because they are located seventeen miles from the main campus. Geographic location may play a role in determining accessibility for disgruntled groups, but it is certainly no test for ascertaining academic accountability. Among university-affiliated Federal Contract Research Centers, Lincoln's budget is second only to that of the Jet Propulsion Laboratory. To restate a point made in Chapter 1: its project funding comes from the DOD (Air Force branch), not in the form of grants but, as with JPL and Rand, in the form of contractual obligations, and most is used to underwrite applied research tasks. To be sure, production engineering is now considered unacceptable there, but its expertise in developing satellite communications, nuclear test monitoring, and early warning radar devices has been well documented. Offensive weaponry these may not be, but national security systems technology of any sort is just as surely not pure research, not expressive activity, and not in any sense an academic species.

What is the larger relationship between the Defense Department and MIT? In 1969, the Lincoln Laboratory budget was comparable to that of

99. Dorothy Nelkin, *The University and Military Research* (Ithaca: Cornell University Press, 1972). The facts presented below come from her account except as otherwise noted.

the entire on-campus research effort. Federal reimbursement for such indirect costs as overhead and administration accruing from Lincoln Lab activities has allowed MIT to allocate far greater resources to academics than would otherwise be possible. Naturally, DOD research subsidies for the main campus cannot compare to the support it funnels to Lincoln, yet MIT receives much more assistance from the Pentagon than does any other institution of higher learning.[100] It is somewhat of an understatement to surmise that the defense interests of the nation and, perforce, the public policies which inform them constitute major configurations in the MIT gestalt.

Industry has also found the institute a comfortable nesting ground.[101] Whereas corporations finance approximately 3.5 percent of academic in-house research, they support 10 percent of MIT's on-campus investigations. The most lucrative agreement is with Exxon, which has been spending $7.5 million over a ten-year period to subsidize combustion studies at the university's Energy Laboratory. The contractual stipulations must send shivers up the spines of local AAUP members. Research areas are mapped out not by MIT faculty solely but by scientists representing both parties, and the company reserves the power not only to prescreen all manuscripts before journal submission but also to insist that publication be withheld for three months so that its patent opportunities will remain inviolate. MIT, meanwhile, has *no* patent rights whatever. When queried about these details, an Energy Laboratory spokesman called the provisions run-of-the-mine.

What, then, is the bottom line? MIT biology is a very expensive research proposition. It costs more than $300,000 a year to support the typical professor's laboratory efforts, while professorial "stars" require $500,000 per annum and sometimes more.[102] Federal funds for education have been attenuating since the expansionary days of the 1960s; the Mansfield Amendment has put a damper on DOD research lacking a national security nexus; President Reagan's budgetary priorities have very much unnerved scientific "managers" throughout the academic world; and gifts from industrialists dabbling in biotechnology do not

100. According to Nelkin, pp. 19 and 126, DOD funding for the MIT campus per se is geared to promote only *basic* research, but she supplies no evidence to support that conclusion. As I suggested in Chapter 1, science in the academic departments there is essentially a pure research–applied research crazy quilt, and any attempt to separate them requires one to distinguish carefully between the sundry commitments of individual centers of study and individual scholars.

101. This paragraph is based on Ann Crittenden, "Industry's Role in Academia," *New York Times,* July 22, 1981, D1, D11.

102. Richard A. Knox, "Secrecy and Questions in Gold-Rush Atmosphere," *Boston Globe,* September 15, 1981, 9.

carry with them the bag and baggage of NIH red tape. MIT was feeling the squeeze and so was MIT gene splicing when Mr. Whitehead arrived on the scene bearing gifts, and just at the time when cloners were beginning to rewrite the textbooks and open wide commercial vistas. The contract which took form could probably never have been written at those private schools which normally guard natural science as a liberal arts component with the greatest zeal, and might never have gotten off the ground even at the far more down-to-earth state universities, which, for all their vocational instincts, must keep their political fences carefully structured. But MIT has never really decided that it wants to be a bona fide university, committed beyond compromise to the pursuit of knowledge and to the political institutions and norms which American education has wrought to preserve the integrity of that pursuit. Rather, its abiding goals have been public welfare and social utility, notions which change color and contour with the ebb and flow of competing ideologies and pressure interests, as Arthur Bentley showed long ago.[103] If the Whitehead-MIT agreement does not comport with tenets central to academic freedom as that notion is commonly practiced, it is not inconsistent with educational norms at MIT, where the marketplace of ideas is constantly hyphenated with a quasi. Perhaps such pluralism in American higher learning is healthy over the long haul; ultimately, it is the scientists themselves who will decide that question when they select the places where they wish to work. And, as I emphasize in Chapter 5, the decision they must make is an overtly political one.

A final word on regulation. As a result of burgeoning academic-industrial relations, Boston, in 1981, enacted an ordinance putting all recombinant DNA research under local control.[104] The law requires every laboratory within city limits to observe NIH guidelines and establishes a biohazards committee, made up of five appointed experts and two elected "residents of neighborhoods," to monitor work at research installations.[105] Both Boston and Cambridge now insist that all institutions conducting cloning investigations apply and *pay for* permits before they can proceed. It is laudable that requests for approval automatically go into effect after ninety days, if not acted upon, and that ordinarily, approved projects will be suspended only subsequent to notice and hearing. It is not so laudable that the commissioner of health and hospitals shall certify the selection of community representatives to sit as members

103. Arthur Bentley, *The Process of Government* (Chicago: University of Chicago Press, 1908). See also Nelkin, *The University and Military Research,* 155–56.

104. "Regulating the Use of Recombinant DNA Technology," Boston Ordinances of 1981, chap. 12, doc. 50-1981.

105. Ibid., 3.

of institutional biosafety committees and that the commissioner also has carte blanche (except as regards trade secrets) to review the minutes of IBC meetings.[106] There can be small doubt that as commercial outfits expand their gene-splicing operations, other grassroots publics will be the more emboldened to pursue their own notions of appropriate regulatory protection, unless Congress decides to preempt the field.

106. Ibid., 5, 8–9.

4 Cloners and Their Watchdogs

From its inception, the recombinant DNA debate has been couched in abstractions, a feature which has not necessarily contributed to elucidation. This chapter focuses on people: the researchers who clone genes and the institutional personnel who monitor the cloning process. We seek to know what kinds of experiments have been commenced and why; we endeavor to ascertain what guidelines are in place and how these are enforced; and we search out the relevant political values and attitudes which inform the scientist as investigator and the academic administrator as watchdog. In short, the analysis here attempts to delineate an operational definition, a constitutional politics overview, of recombinant DNA activity in the American university.

During the fall of 1981, I conducted a series of personal interviews with twenty-six participants in the gene-splicing regulatory process, nineteen cloners, and seven chairmen of campus biosafety committees.[1] Sample selection of at least one scientist from a particular facility automatically triggered inclusion of that facility's IBC panel head on my interview agenda, for one cannot explore institutional research governance as a set of complete enterprises without accounting for the beliefs and actions of both scientist and administrator. Geneticists with principal appointments in government agencies such as the National Institutes of Health, in commercial enterprises, and in nonprofit, nonacademic units were excluded because I was attempting, first and foremost, to capture the essence of scientific analysis (in this case, of recombinant DNA experimentation) as a species of free expression. Those who work for NIH can be held, under my theory of constitutional right, to a far higher standard of accountability than can the college professor, and it would be

1. Appendix A describes the procedures employed to isolate both sets of respondents.

111

counterproductive to provide them with representation here. Corporate research environments were eliminated for the obvious reason that the NIH guidelines are not binding upon their scientists. Finally, I put to one side investigations sponsored at institutions such as Cold Spring Harbor, not because there is any doubt as to their prima facie standing under First Amendment doctrine, but because the term "academic freedom" has special relevance to degree-granting institutions of learning, and it seemed best, in this first study of its kind, to deal with a category sufficiently homogeneous to optimize the testing of applicable theory. For logistical purposes, the sample of interviewees was drawn from institutions located in the Boston-Washington corridor. I do not believe that this geographic concentration has caused any bias in my findings; indeed, as I indicate in Appendix A, the nation's northeast quadrant is one of the two territorial "heartbeats" of cloning experimentation in the United States.

The discussion which follows contains many references to particular questionnaire items. In order to make reliable and valid comparisons among respondents' observations, I employed two master questionnaires, one geared to the scientist, the other geared to the IBC supervisor.[2] Some items, of course, were tendered to members of both groups, allowing the opportunity to test for differential perceptions of the same phenomenon. The point is that social scientists are not media reporters; they aim to standardize stimuli so as to neutralize intervening factors which might prejudice results. On the other hand, there may be circumstances when greater flexibility should control. If respondents felt compelled to stray from the mark, as researchers define that mark, and discuss other issues which may prove to be of even greater pertinence, it was my practice to let them talk and even to encourage such dialogue with on-the-spot queries. But always I returned to the prepared instrument when the subject seemed out of play. Each questionnaire included approximately forty-five items, and interviews ran anywhere from 40 to 120 minutes. We begin with the recombinant DNA scientist.

Research Designs: Names, Ranks, Serial Numbers

Our first task is to conceptualize the nature of principal investigators' cloning exercises: their purposes, their significance, their funding, their constitutional standing. That process commences by addressing the specific objectives of respondents' gene-splicing studies. After all, we cannot analyze the impact of administrative guidelines if we do not understand the scientific enterprise at which the guidelines are being di-

---

2. Appendix B presents the questionnaire administered to recombinant DNA investigators, and Appendix C provides the roster used to interview committee chairmen.

rected. And we cannot classify experimentation as "expressive activity" if we do not know what the research mission is designed to accomplish.

It is not an easy task to generalize about interviewees' discrete interests, but certain common themes do emerge. By far the most prevalent is "gene expression"; that is, a clear majority of projects were geared toward understanding how particular genes ultimately produce particular proteins. This group, moreover, can be divided into three smaller clusters: one subset was devoted to analyzing mammalian genetic behavior; a second emphasized the hereditary functions of other eukaryotes and a third concerned itself with understanding DNA expression in viruses.[3] An instance of the first category was a project aimed at investigating the expression of globin genes in humans. A good example of the second was a study of genetic regulation in Drosophila (fruit flies). The third subset is exemplified by an attempt to determine the activity of certain genetic material in the Epstein-Barr virus. A second common denominator is viral gene construction. "Here we have an organism that causes disease," one researcher said; "How is such-and-such hereditary information within that creature organized?" Still another investigator was working to decipher part of the DNA code language for the feared herpes virus. Genetic sequencing of nonviral species was also cited. Not surprisingly, there is some overlap in the data, as several practitioners said they were interested in both structural and behavioral topics.

To recombinant DNA scientists, contriving mutants is a "tool," a "technique," a "procedure," a "mode of analysis," a "gene isolator," and a "gene purifier." In other words, gene splicing plays the same role in their endeavors as the microscope, the telescope, and the X-ray crystallographic procedure play in the execution of other research probes. So cloning is perceived here simply as a means to an end. But we cannot accept that wisdom at face value. To utilize a microscope is to sharpen one's vision, not to effect a decisive change in the subject species' makeup. And while such processes as freezing and heating can also constitute research methodologies—working significant alterations in the matter to which they are applied—neither create what has never before existed in nature and what is now to be classified as a life form. So tools come in many descriptions, depending greatly upon the results they accomplish. Actually, a handful of respondents indicated that at least some of their experiments were indeed innovative. One remarked that his clones were a form of artwork, while another called his plasmid hybrids a "creative enterprise" which, he hoped, would expand the horizons of recombinant DNA research as a substantive discipline. But most scientists in my sample did not define their work in such fashion, and while

3. A lone enterprise addressed genetic functions in prokaryotes.

this may well reflect the nature of the typical gene-splicing labor, a presumption one must now indulge, I would hypothesize that there is a clear gap between the scientist's perception of cloning as, generally speaking, a workaday enterprise and the informed public's perception of cloning as an exotic, unique, grand strategy.

In most cases, scientists felt their larger research goals could not be achieved without cloning, and in the other instances, respondents believed meaningful results would be exceedingly more difficult to attain if they had to rely on tissue cultures and other, more conventional, instruments. The notion of recombinant DNA methodologies as indispensable for their needs captures the overwhelming support of both the large group that clones to achieve an end and the small group that clones seemingly as an end in itself.

Is the science we have been discussing pure or applied research? Two tests, one objective, the other subjective, are available to us. With the former, decisionmakers must render an independent judgment based on the totality of facts presented. For example, where the dominant theme, taken as a whole, is the investigation of nucleotide sequence or amino acid arrangement, we have little trouble concluding that the endeavor is basic research. As our universe of study is the academic laboratory, we would expect the typical project to be one of abstract analysis, and indeed, no fewer that eighteen blueprints (out of twenty-three) satisfy this criterion, beyond any shadow of doubt. With the latter, subjective model, participants judge their own work, employing whatever standards they consider appropriate. The flaw in this approach is the increased possibility of personal bias, but the test has value, especially as a counterweight to the opinion of nonexperts. While judicial officers, say, could seek the advice of knowledgeable scientific witnesses when necessary, I am willing, within the confines of this particular scholarly presentation, to compare my views with those of the principal investigators themselves. Respecting seventeen of the eighteen studies I rated as "pure" beyond cavil, the geneticists involved reported, independently, the same conclusion. The deviant instance involved the very same subject which raised cautions regarding the other five investigations, namely, situations where hospital patients would be recipients of DNA-engineered materials. In this particular project, the major thrust of the endeavor appeared to be a determination as to whether two paramyxo viruses cause multiple sclerosis, but the researcher, taking a more conservative view, emphasized the fact that he had been working especially hard at developing "DNA probes," which he evidently considered a sort of "viral package" or diagnostic product.

As to the five research designs which seemed, on their face, offbeat as academic enterprises, subjective and objective analyses proceeded as follows: (1) A respondent wanted to study sequencing, describe protein expression, and construct artifacts to make better vaccines. Here the totality seemed weighted toward pure science, especially as preparations for medical manufacture had not yet commenced, but the investigator labeled his efforts "pure and applied," stressing the medical orientation of the exploration. (2) An investigator was attempting to understand why certain tumor cells in patients were drug resistant and was developing a recombinant probe to examine molecular changes. In this case neither subjective nor objective assessment could resolve the competing variables. As the interviewee put it, the experiments had never before been done, yet the benefits that might accrue to the patient would be clear and present. (3) The purpose was to mass-produce a vaccine so as to assist poor people in Third World countries suffering from various diseases brought on by parasitic invasion. Both standards of review yielded an applied research characterization. (4 and 5) Again, the avowed goal was gene therapy, and subjective and objective judgments once more concurred: viz., if the research has directed application to specific clinical entities and if steps have been taken to prepare unique tools for the purpose of fulfilling the goals of that endeavor, then the investigations are clearly not exercises in basic science.

The issues we confront at this level of inquiry are crucial, because upon their resolution depends the manner in which scientific activity triggers free expression coverage. I argue, first, that my methodologies for ascertaining the fundamental nature and purposes of DNA recombination were essentially sound, since in regard to 91 percent of the studies subjected to content analysis, writer and researcher agreed in their conclusions, without even stipulating the ground rules of investigation. It would appear that criteria for judgment are fairly well developed within the academic community, though no jurisprudence exists on the matter. I submit further that 73 percent (17 our of 23) of these experimental designs are pure science and merit classification as quasi-speech "ways of knowing." Because gene splicing is evidently necessary to the success of thirteen of those seventeen projects, I believe that regulations proscribing the procedure under these circumstances would be unconstitutional, unless some extraordinary burden of proof were satisfied. Three investigative operations, I conclude, are applied science and deserve no especial constitutional solicitude. Finally, there is a twilight-zone cluster of cloning efforts which require further sharpening before they can be placed on the pure-applied continuum, before they can be defined as quasi speech or mere conduct. These impediments could readily be overcome if knowledgeable parties agreed that the "totality" test is the best available calculus by which to classify experimental research, and if they also agreed that genetic therapy, as a specific and immediate applied activity, carries more weight in that calculus than do the basic research components of any particular cloned delivery system. The strength of the totality notion is that it provides fullest allowance for context and intention, as attempts to define "due process" and "cruel and unusual punishments" graphically exhibit. The strength of resolving all doubts as to quasi speech standing against overt attempts to help the lot of patients is that it comports with our great respect for human dignity in the face of affliction, a concern which should accord to the state the freest possible regulatory hand.

An interesting footnote to researcher self-evaluation is that there is a tendency for those who view their recombinant DNA investigations as creative to also view them as applied research! That is, the scientists who strive to develop the new clone, the exciting hybrid, are, in this sample, investigators who want to find the cure for particular diseases and demonstrate appropriate proofs thereof. For them, the ideal clone is an *invention*, a technological breakthrough which will help solve the problems of humankind. Presumably, this is the spirit of the patent grant in modern dress, where the goals of science and useful arts merge in the fruits of discovery. But, as before, I find no First Amendment interest to protect. In any event, the numbers in our sample are so few in this regard that these observations must be considered only suggestive.

Regarding the nature and scope of their funding sources, scientists report, as anticipated, that the National Institutes of Health are, by far, their chief benefactor. More precisely, 79 percent receive money from Bethesda, but only half rely solely on NIH for support. The others derive supplemental assistance, some of it substantial, from a variety of private agencies, the most generous being the American Cancer Society. The National Science Foundation lags farther behind than even the data in Chapter 1 might have predicted, subsidizing only one gene splicer in toto and providing partial aid to two others. With respect to numbers, respondents considered most accessible their per annum, direct cost, project award emoluments, figures which are published in *NIH Research Grants* for those funded out of Bethesda's pockets.[4] It is upon these estimates that I rely. The cloning activities indulged in by my sample cost, for one year, $2,131,000, and the average *total* award per recipient is $112,160. It is surprising that these figures are so low because the typical NIH support package, not even including assistance from others,

4. For instance, *National Institutes of Health—Research Grants,* NIH Pub. No. 81-1042, provides outlays to grant holders for FY 1980.

Table 4.1. Classification of respondents' experiments according to NIH guidelines

Guideline requirement	Number of studies	
	Original evaluation	Revised evaluation
Prohibited	2	0
P3-EK2	2	0
P2-EK2	4	0
P2-EK1	4	0
P1-EK2	0	4
P2-EK0	1	0
P1-EK1	6	10
P1-EK0	0	1
Exempt	0	5
	19	20

is $124,000.[5] Two investigators reported gifts of $250,000, while four had stipends valued at $200,000. These highs were counterbalanced by one award of only $25,000 and the unfunded research of another investigator, which was priced at $5,000. Again, NSF's contribution is extraordinarily weak, for according to my rough calculations, the National Science Foundation did not supply even 10 percent of the total package. Lastly, both those who evaluate their cloning as applied or quasi-applied research and those who rate their cloning efforts as not indispensable to their long-term objectives are scattered up and down the reward system totem pole.

The NIH Guidelines: A View From the Laboratory

All nineteen scholars stated that their work had, at some time, come within the purview of NIH supervisory jurisdiction. For the small group funded solely from other sources, the guidelines were made applicable either because the grant-givers—e.g., NSF and USDA—had adopted NIH rules for their own or because the university/employer received NIH monies for other cloning purposes. These constituted the only formal restrictions binding upon respondents, as the sample failed to include any Cambridge researchers and the Boston Biohazards Committee had not yet held its first meeting.

Table 4.1 indicates the guidelines initially imposed on these efforts and the extent to which they had undergone revision. The second column

5. Marjorie Sun, "Researchers Predict Fewer NIH Grants," *Science* 215 (March 26, 1982): 1599.

gives a breakdown for nineteen studies[6] as NIH watchdogs first evaluated them, running from most to least severe constraints, while the third column shows the prevailing requirements as of November 1981 for these same experimental designs plus one experiment recently commenced. These data indicate that P1-EK1 was and is the most frequent designation, but whereas P2 standards were the mode level of usage early on, all such classification orders have now been revised. I was fortunate in that the campus biosafety panels with which I worked, unlike ORDA, kept complete records of gene-splicing research in their communities, else I would not have learned that 25 percent of these studies are now exempt under the NIH guidelines. That figure, moreover, seems unexpectedly low, given Bethesda's claims as to how much cloning it *thinks* is no longer covered. All told, 80 percent of projects for which I have information have undergone a process of deregulation.

There seems to be no correlation between the stringency of guidelines and the purposes of the studies to which they have been assigned. The prime consideration has always been what sort of host, what sort of vector, what sort of hybrid is being employed. The fact that experiments may be exercises in basic as opposed to applied science has never been deemed salient in determining the rigor of the administrative restriction.

Generally speaking, the recombinant DNA scientists interviewed have supported the particular guidelines pegged to their work, but there is substantial minority sentiment critical of NIH's judgments and some of that opinion can only be called strident. When asked what physical and biological containment they would place on their own experiments, respondents rated 62 percent of their studies as Bethesda rated them. Yet there were eight dissenting evaluations, and surprisingly, three of these thought NIH was too permissive. Thus, two cloners whose projects were exempt said that "standards of the profession" dictated the use of P1-EK1, and they thought the regulations should prescribe these as minimum conditions. In fact, as much as one-third of all gene-splicing activities (7 out of 21) were being conducted under precautions more conservative than those mandated by NIH. For instance, four out of five exempted projects have been developed with EK1 specimens. On the other hand, three geneticists held to P1-EK1 standards thought it unwarranted to put any criteria on their work, which they regarded as eminently safe. But are these simply issues on which reasonable people can differ? According to a somewhat different minority of seven (out of 21

6. Generally speaking, I code "respondents," not "studies." However, as with questions 1 and 2, principal investigators sometimes note that different conditions obtain for different projects. When interviewees make these distinctions, I talk of research efforts rather than the scientists who direct them.

evaluations), the answer to that question is no. In other words, NIH's judgment has not only been questionable; on the whole it has also been unreasonable, arbitrary, and capricious with respect to these respondents' research prerogatives. Question 9 was framed deliberately to test for this low threshold of endorsement, because I believe "reasonableness" to embody the prevailing *constitutional* criterion for "times, places, and manner" strings affixed by government agencies to the funding of quasi speech. And I think the comments elicited demonstrate that the researchers well understood the nature of the inquiry. Hence, the two investigators who felt NIH should have placed P1-EK1 standards on their exempted work said the agency's determinations were reasonable but imprudent. On the other side were two scientists who believed it indefensible that their work had once been prohibited, another who thought the P2 levels earlier assigned to his investigations nonsensical, and a third who felt the guidelines were so absurd they should be abolished forthwith, not because they were too tough but because his research—hybridizing human and SV40 DNA—was rated P1 rather than the P3 level he had been using on his own initiative. The underlying thread running through these minority convictions is that NIH has been motivated largely by political rather than by scientific considerations and that Bethesda's track record has not comported with the state of the professional art.

I asked scientists to describe the role of the campus biosafety committees in monitoring their experiments and learned that one out of two laboratory facilities had never even been examined prior to the onset of gene-splicing activities, much less subjected to on-the-spot inspections. Some of the remarks by principal investigators border on the derisive: "They never look us over, but they wouldn't know a violation if they saw one." "It's P1 work, so nobody cares." "I don't know they exist. Who's the Chairman?" "The biosafety officer asked me if my clones were hazardous. This shows the researcher has total control. After all, we have the expertise." In some instances, a committee representative did go through the premises, checking the equipment, posting signs, and asking how waste products would be processed. Intensive committee reviews, complete with instructions on the proper handling of volatile materials and the correct safety uniforms to wear, were infrequent occurrences even among those doing "sensitive" (P2 or higher) work. It is little wonder, then, that none of these DNA practitioners could point to a bone of contention having arisen between themselves and oversight personnel. In fact, considerable praise was heaped on one panel chairman because of his understanding and cooperation. There was some small criticism, however, of the rule which allows gene splicers to sit on review committees and, in a sense, pass judgment not only on others but on their own work ways. Not even "guidelines," it was argued, should be implemented this way.

In Chapter 2, I expressed concern about the process whereby cloners were obliged to receive clearance from campus authorities whenever they wished to depart from the letter of their research protocols. It turns out that only about a fifth of our sample had ever submitted such petitions for review, that the issue of timeliness had raised no problems because biosafety officials never took more than two months in processing these requests, and that the standard of "reasonableness" as defined by gene splicers themselves was satisfied without exception for this aspect of NIH supervision. Again, however, a note of cynicism emerges, as some scientists deliberately contrived their project designs in language capacious enough so as to avoid any such "bureaucratic hangups."

Cloners believe—unanimously—that biosafety committee guideline *interpretations,* in contradistinction to the letter of the guidelines themselves, were consistently within the zone of arguable value judgment. I found no evidence to support an allegation that these panels, by their nature and in the performance of their duties, are authoritarian or excessive in their demands. If criticisms are to be lodged, they involve, first, a spirit of informality pervading the guideline implementation process and, second, the broader issues of committee competence and Bethesda's rule-making determinations, which a vocal minority thinks verge on the arbitrary.

Even deeper divisions surface when recombinant DNA workers address the total context of federal restrictions as functional norms. Hence, only a bare majority thought government supervision of their research, taken as a whole, evinced a commendable sense of proportion. Not that oversight groups had offended anyone by being either too active or too passive. Rather, several investigators argued that the level of supervision was acceptable only because there was *no* supervision for all intents and purposes! Still other critics thought NIH had done a poor job of setting up the entire regimen. As these cloners saw things, Bethesda had given the committees carte blanche, had never developed a strategy of compliance, and, reiterating a theme sounded earlier, had vacillated in policy orientation between the "overkill" of 1976 and the lassitude of 1981. The same sort of ambivalence surfaced when investigators were asked if government watchdogging was efficient or inefficient. Again, a slim majority gave the forces of oversight a passing mark, but almost half of these scholars thought national headquarters was totally out of the picture, a phenomenon they regarded with approval. Some members of the panel said they simply didn't know, stating that because they never interacted with their committees or felt themselves subject to supervision, they were unable to judge the efficiency of the regulatory process. And an equal number went so far as to call the entire guideline enforcement mechanism inefficient, striking out against bureaucratic stupidity, confusion, and paperwork.

If these data have larger significance, it would appear that agreement among recombinant DNA investigators regarding the merits of the government/administrative supervision accorded their experiments hangs by a thread. But there is no substantial, ideologically oriented dissenting faction. For example, of the scientists who said their campus committees did practically nothing by way of monitoring, about half reported that government oversight, taken as a total package, was about right and two-thirds stated that such oversight was either efficient or that they could not form an intelligent appraisal. So there seems to be a generalized feeling that the enforcement process lacks coherence, predictability, and follow-through. Respondents want guidelines, but many either heave a sigh of relief when the restrictions are given only lip-service or lash out at the hypocrisy they see when rules are promulgated but not enforced diligently. And, as already noted, several members of the sample do question the rules that have been fashioned to fit particular cases.

For someone sitting in ORDA headquarters, it must appear as though Bethesda is being placed in a sort of "damned if we do, damned if we don't" dilemma, and a recurring question is whether NIH has only itself to blame. Good politics, within the framework of constitutional republicanism, is very often nothing more than coercive authority channeled through legitimate, rational norm structures. Whatever these terms may mean at other times and places, it seems clear to many scientists working in this area that NIH has never been able to vest in the guidelines a viable political capability. The upshot is that while there continues to be a broad consensus on general principles of restraint, each community—indeed, each laboratory—has its own story to tell of professional ambition tempered with disillusionment over "great debate" politics. And these commentaries, taken together, make one wonder if the guidelines as *national policy* occupy a salient niche in the belief systems of more than one out of every two gene cloners.

After all is said and done, do principal investigators accept the guidelines spawned at Asilomar? We have seen that a majority favors the sorts of guidelines that Bethesda has been proclaiming, but that by no means proves that gene splicers want NIH to continue to lay down any set of norms as *mandatory* constraints. An overwhelming body of opinion (74%) thinks NIH should retain obligatory rules. The following reasons predominate: "Some of this work is hazardous, and the guidelines keep people honest." "RAC has expertise in this area that should guide our efforts." "If NIH moves out, the city councils will move in, and the result would be anarchy and stupidity." The minority provided these views: "The guidelines have outlived their usefulness. Recombinant DNA science is not unique." "If we got rid of the guidelines, it would stop all the fakery. There is a ton of cheating, and nobody does anything about it." "NIH is not a regulatory agency, and the guidelines are nothing but regulations." To sound a previous note, I wish to emphasize the weak relationship between this dissenting bloc and others I have identified. For example, only 60 percent of those who would make the rules voluntary contended that these standards were unreasonable. It also turns out that 80 percent of those who object to compulsory national controls also object to compulsory campus controls. So, perhaps not unexpectedly, we have some sentiment for the proposition that investigators ought to be left to their own devices and that all forms of administrative restrictions are unjustifiable impositions on research autonomy.

There was considerable question whether, in retrospect, cloners believed the Asilomar majority had acted wisely in lobbying NIH for mandatory recombinant DNA constraints. I believed that some researchers might accept the status quo because it was the best of all possible worlds, though if they had a chance to do it all over again in the light of a more informed political consciousness, they would resist the temptation to open what they now regard as a Pandora's box. I found, instead, practically no inclination to second guessing: 84 percent asserted that the Asilomar-NIH initiatives have constituted the proper political framework within which scientists should work to develop cloning as a mature discipline. Why? These comments tell the story: "If we hadn't acted, Congress would have passed a law depriving us of rights." "It took time to show that the research wasn't as hazardous as some people feared." "The restrictions slowed down the flow of knowledge, but it was good politics to show people we had a social conscience." "Some of the work is dangerous, and we needed constraints."

So principal investigators, with all their sundry misgivings about guideline implementation, approve, virtually without equivocation, both the letter and the spirit of a national rule structure circumscribing their research. We have lived for the past fifty years in an expanding universe of federal police power and responsibility, but when the coercive shoe pinches, or squeaks, or begins to wear thin, we have not been above complaining that the idea is good but that human behavior is getting in the way. Eventually (Chapter 5) we must ask whether we have here just another example of wanting what constitutional arrangements were never meant to accomplish. But now it is appropriate to examine those delivery systems, not from the standpoint of persons being held to account, but from the vantage point of persons commissioned to govern.

Institutional Biosafety Committees: Names, Ranks, Serial Numbers

The next several sections describe the organizational dynamics and the procedural norm structures displayed by NIH recombinant DNA biosafety panels, as seen through the eyes of seven committee chairmen. Where appropriate, I compare my findings with those reported in Diana Dutton's 1979 study of California IBCs.[7] That project had as its chief aim an analysis of public participation in the monitoring of biomedical research, a goal which is of minor concern here. I should also point out other differences in our investigations: Dutton interviewed three times the number of committee heads to whom I spoke, providing her with a clearly superior universe; on the other hand, she employed a mailed questionnaire instrument, and my experience with this form of interrogation is that it often provides respondents with a license to frame perfunctory, sometimes irrelevant, replies. Finally, I characterize various opinion data and perceptions of the guideline enforcement process with which she had little concern. That there is substantial agreement between the demographics reported in our surveys, however, certainly indicates face validity for the procedures employed here.

It is important to emphasize, right at the outset, that institutional biosafety committee chairmen, unlike the recombinant DNA scientific community as a whole, have played a direct and significant part in molding federal cloning policy. In November of 1980, representatives of 154 IBCs from around the country convened for a series of brainstorming sessions on matters of mutual interest. At this meeting, the panel heads cast two straw votes of far-reaching importance. First, they overwhelmingly opposed a plan then under consideration at NIH headquarters to work up a

7. Dutton's research is described in U.S. Department of Health, Education, and Welfare, National Institutes of Health, *Recombinant DNA Research: Documents Relating to "NIH Guidelines for Research Involving Recombinant DNA Molecules,"* 6 (April 1981): 162–63, 199–202. Hereinafter referred to as *Recombinant DNA Research*. Further information on IBC performance can be found in Philip L. Bereano, "Institutional Biosafety Committees and the Inadequacies of Risk Regulation," *Science, Technology, and Human Values* 9 (Fall 1984): 16–34. For an interview schedule employed in questioning the chairmen of institutional review boards, see Bernard Barber et al., *Research on Human Subjects* (New York: Sage, 1973), 200–232, especially 205–14. Cf. Bradford H. Gray et al., "Research Involving Human Subjects," *Science* 201 (September 22, 1978): 1094–1101.

formal analysis and evaluation of IBC functions. That reaction cooled considerably Bethesda's ardor for the proposal, and it was shortly placed on the back burner. The result is that we have never had an in-depth national study of recombinant DNA campus oversight; prior to my analysis, only the Dutton study had provided any insights since the 1978 deregulations. Secondly, the chairmen in attendance recommended that virtually all nonprohibited experiments employing *E. coli* K-12 host-vector systems be exempted from the guidelines. These projects, they thought, were not only eminently safe but also generated by far the greatest amount of paperwork for which their committees were responsible. At the RAC meeting of April 1981, much sentiment was expressed sympathetic to these administrative burdens, and by a tally of 13 to 8 a motion calling for such exemption was enacted, amended only by stipulations that P1 controls be recommended (though not required) and that large-scale work of this nature remain under the guidelines—that is, retain their prohibited status—unless exceptions were obtained in the conventional fashion. So great was the momentum for deregulation that RAC also eliminated both the provision for IBC prior review of *E. coli* K-12 cloning designed to express eukaryotic protein genes and the guideline coverage theretofore extended to the vast majority of *B. subtilis* host-vector work. These actions Bethesda approved without reservation.[8] Clearly, NIH campus personnel have had no compunctions about voicing their opinions on the recombinant DNA debate when the spirit moves them. The question then becomes what roles they perform in the day-to-day events of that debate.

The average number of members sitting on the IBCs that I studied was fourteen, ranging from a board as large as twenty-five to one with seven panelists. Dutton found California agencies to be somewhat smaller, with the typical unit having eleven participants. A unique division of labor features the workings of the twenty-five–person body, which has delegated the disposition of biohazards questions, including those involving review of recombinant DNA issues, to a six-member subcommittee, some of whom do not even sit on the parent IBC. It is important to note that the *larger* unit is the institutional biosafety committee, because the two public representatives mandated under NIH cloning restrictions belong to it, not to the minipanel. Needless to say, this power-sharing administrative mechanism is quite likely in violation of Bethesda's intentions, if not stated policy, because public participation in recombinant DNA assessment seems clearly to mean something substantially more

8. *Fed. Reg.* 46 (July 1, 1981), 34454–55. The November 1980 meeting elaborated upon above is discussed in Elizabeth Milewski, ''Report of the IBC Chairpersons' Meeting,'' *Recombinant DNA Technical Bulletin* 4 (April 1981): 19–27.

Table 4.2. Classification of respondent IBC membership by professional qualification

	Total in all panels	Average per panel	California average (Dutton study)
Cloners	23	3.3	3.0
Other scientists	34	4.9	3.1
Social scientists	1	0.1	—
Physicians	11	1.6	—
Lawyers	5	0.7	—
Administrators	8	1.1	2.4
Community representatives	13	1.9	2.6
Students	3	0.4	0.3
Others	2	0.3	—
All categories	79	11.0	10.8

meaningful than having a voice when state occasions arise on appeal. In Table 4.2, I provide a series of data respecting the professional credentials of biosafety personnel; as it is the six-person subcommittee that actually processes gene-splicing projects for the one particular campus, the information cited is based on the composition of this smaller forum.

Our typical committee has 11 members, and 3.3 of them are recombinant DNA specialists, figures almost exactly in accord with those Dutton found. My study showed a slightly larger representation for other natural and physical scientists, a slightly smaller representation for administrators, virtually an identical figure for student membership, and a disparity in community representation which is caused entirely by the "violation" discussed above.[9] Social scientists, humanists, and lawyers appear to play little role in institutional oversight. As one IBC chairman explained, "When it became clear that the committee would make largely technical decisions, we felt no need to include these people and, from what I can tell, they are happy not to be involved." Moreover, several respondents argued that the 20 percent allotment for outsiders has made it necessary to cut back on overall committee size. "If you have twenty members, then four must represent the community. It's hard to find that many who both have expertise and the willingness to meet every month. So we have reduced our size, and because we must have good scientific know-how,

9. As Dutton used somewhat different questions than I did, a few of the percentages imputed to the California scene come from aggregations of her data. Thus, I combined her categories 2 and 3 to obtain a score for "other scientists" and brought together her categories 5, 6, and 7 to determine the participation of administrators. See *Recombinant DNA Research* 6 (April 1981): 199. Dutton's survey also included nonacademic units, and I cannot determine what differences that would make.

the others sort of opt out." Most of the community participants in my sample were either recombinant DNA scientists in their own right or possessed other expertise relevant to biohazards questions. In Dutton's study, these select groups also fulfilled that function.[10] While I made no special effort to ascertain institutional reaction to the 20 percent outsider rule, unsolicited commentary from IBC leadership makes me doubt that a solid majority considers the provision appropriate, a conclusion which is contrary to the California findings.[11]

Biosafety committee chairmen were divided on whether they thought they possessed special skills or expertise respecting panel responsibilities. Three said they did, one citing the fact that he was an original member of RAC, a second pointing to his status as a recombinant DNA principal investigator, and a third, the only one of the seven panel chairmen who was not a natural scientist, noting he had been affiliated with the Hastings Center. The others felt they lacked any distinguishing characteristics, reporting only their academic specialties and the fact that they had attended the meeting of IBC chairmen convened by NIH in November 1980. Our panel heads were also divided as to whether they performed any special tasks in the administrative process, aside from presiding over meetings. Three interviewees felt they did little else, but four were more active, remarking, for example, that they screened all proposals, decided if documents should be handled through the mails, over the telephone, or by a full-blown committee review, and even made interim judgments as to containment levels in order that experimentation could proceed before the next scheduled meeting. Whereas Dutton found that review board members in California often received special training regarding cloning endeavors, the panels I studied underwent little to no formal preparation. Some of these committees, however, had the benefit of a biosafety officer in residence, as NIH requires that such person be appointed and delegated special tasks wherever P3 work is done. Still, there exists no tendency among panel chairmen to dominate, either via the impress of credentials or the impress of delegated authority, the IBC decisionmaking process.

A Proposal Becomes a Project

Once principal investigators have apprised appropriate institutional biosafety committees of their intentions to commence cloning experiments, how do campus officials process that information? We must distinguish between the old system, which required the parties to work up memo-

10. Among her 2.6 "nonaffiliated" panelists, only 0.8 were "local citizens," or 31%. Ibid., 199, 201.
11. Ibid., 163.

randa of understanding, and the new system, wherein most notifications are largely in the form of registration papers. Dutton's data—gathered under the former rules of the game—show that 73 percent of proposals were accepted without modification.[12] My findings are strikingly similar: IBC chairmen said that 76 percent of the memoranda petitions filed emerged unscathed after panel deliberations. One committee head reported that his board had never turned back an initial application, but two others said that as many as 50 percent necessitated more work. I anticipated a dramatic decrease in rejections under the new registration system, but the 76 percent approval rate rose only to 82 percent, and the range continued to hold as before. The problem, as summed up by one IBC chairman, seems to be that "scientists don't know anything about safety matters." Another said: "At first principal investigators weren't familiar with the rules. They are more informed now, but many still don't do the necessary homework and you just have to ask for more information." And there was the matter of the guidelines themselves. "They were so complicated, yet so ambiguous. All of us were confused." However, the general sentiment was that while cloning practitioners have done fairly well in developing plans of action and protection, they would—and should—have done better but for their single-minded research concerns. Two committees make this "first-screen" function the chairman's prerogative; that is, he rejects on his own authority proposals he considers improper and recommends requisite alterations. That convention obtains for precisely the panels where the head appears to have the most expertise.

With their proposals having been returned to the drawing board, gene splicers are almost always able to mount satisfactory improvements, generally by increasing the level of security and sophistication of safety equipment in the workplace. Three of the panels have eventually accepted every set of documents submitted for approval, and two others report a success rate on reruns of over 95 percent. Very likely, Dutton's 4 percent permanent rejection figure is higher than the comparable figure for my test group;[13] perhaps the deregulations which occurred between 1979 and 1981 make the difference. Why would a proposal never receive the green light? The typical instance involved scientists who thought they would be embarking on P2 experiments, only to learn that their work was classified P3, a level of containment for which the institution had no facilities.

An interesting due process question is whether campus IBCs provide a

12. Ibid., 200.
13. Ibid. One IBC spokesman noted that 50% of petitions never obtained the go-ahead, but the translation in raw numbers is one out of two.

mechanism of appellate review in the event that recombinant DNA researchers think they have not received satisfaction in these exchanges. All the panel chairmen agreed that appeals could be taken, but there was some confusion as to who would handle them. A majority indicated that an "upstairs" administrator could step in, the most popular choice being the person who had appointed the IBC membership. Two respondents said the disgruntled party would have to contact Bethesda. In the case of the campus with a biosafety-biohazards division of committee labor, appeals would be permitted from the chairman's findings to the six-person subcommittee but could not proceed to the twenty-five–person full panel. None of these committees, however, reports a single appeal having been lodged against their determinations. As one panel head remarked: "We are a collegial enterprise; we are not adversaries. We try to point out problems, we would listen to any new evidence, and, of course, we are always willing to negotiate if there is any room for maneuverability. When we have been forced to nip a proposal in the bud, our colleagues take their medicine."

When a research proposal is being considered, most IBCs permit principal investigators to make presentations to the committee regarding the nature of their work but ask them to leave when the floor is thrown open to discussion or when votes must be taken. Researchers generally avail themselves of this opportunity, and any hint of trouble is always sufficient inducement. Evidently, there is some difference of opinion in the typical panel's deliberations, though considerably less than during the late 1970s when much confusion existed regarding the meaning of NIH instructions. But it is not possible to generalize about the presence and nature of factionalism on these panels, viz., whether the losing side in the usual controversy over regulatory standards tends to endorse greater or lesser levels of constraint. Two IBC chairmen rated their majority blocs as tilted very much in the direction of lesser restrictions; two others saw their panels as dominated by forces advocating more stringent controls. Said one spokesman: "We think the rules go too far, and only a few among us disagree." But another respondent countered: "The guidelines represent minimum standards of uniformity; they are a foundation upon which to build safer research within our community." And then there was this observation: "At first, our guard was up. Those who thought the recombinant DNA debate was a tempest in a teapot were on the defensive. Now we sometimes ask ourselves why we are still in business." These data provide considerable indication that committee watchdogs are as contentious over the root issues posed by the NIH guidelines as are the cloners themselves.

Finally, I sought to determine how these dialogues are translated into IBC policy. Panel chairmen say that their committees, without exception, make decisions by formal vote. A scant majority keeps a record of such tallies, but in a few committees where no compilations are attempted, all votes have been unanimous, whether positive or negative; in that sense, then, reliable accountings do exist. The question of what constituted a winning margin reaped unexpectedly rewarding insights. In four cases, motions carry by majority or plurality division, but three panels employ versions of Calhoun's unanimity principle. For example, on one IBC, if a member casts a negative ballot, then the committee automatically holds the proposal up until the principal investigator provides more details or better evidence. On another panel, a divided vote has never occurred, but if it did, the chairman said all doubts would be resolved in favor of the wavering or negative sentiment. Even one of the review groups utilizing majoritarian parliamentary procedure would probably be swayed by strong minority objection, its chairman informed me. "We count hands," he stated, "only when we know all doubts have been resolved." Then there is the committee which doesn't see the need to vote much anymore. "Now that people are required only to register, we rarely have occasion to deliberate and make these sorts of decisions." Dutton reported that, in her sample, 47 percent of decisions were by majority and 42 percent by consensus.[14] For our group, this percentage variation is very probably accurate, except that "by consensus" should be amended to "by concurrent majority." The point is that the collegial spirit can be eminently flexible, but it can also accord veto power to a strident few. And in this context, where all committee action is precipitated by the inquisitive researcher, the Calhoun model emerges as a tool that can be utilized only to slow down the momentum for scientific investigation.

Project Oversight and Guidelines Overview

Once IBC clearance has been achieved, can researchers proceed with a free hand? How are NIH standards implemented following assignment? Does Bethesda play any meaningful role in committee enforcement policy? How do IBCs interact with their parent institutions in the context of recombinant DNA politics, and how do they evaluate their place, and the place of the guidelines, in the total context of academic research interests? These are some of the questions we examine in this section.

We begin by obtaining committee perspective on the process which permits (obliges?) principal investigators to renegotiate their research protocols before they perform unauthorized experiments. The procedure has been invoked only on infrequent occasions, by which is meant less than a half-dozen instances per panel. When asked what happened to these initiatives, no spokesman mentioned a single case of refusal, and

14. Ibid., 201.

the majority was certain that every request for a change in research strategy had been granted. On the issue of timeliness, the typical IBC processed forms in approximately twenty-five days, the range being from two to five weeks. I am satisfied, as were the practitioners to whom I talked, that committee renegotiation procedures have been consistent with standards of reasonableness.

On the question of whether NIH national headquarters involves itself in campus recombinant DNA affairs, a healthy majority of chairmen said they had engaged ORDA staff in telephone conversations, attempting to ascertain how the guidelines should be applied in particular circumstances, and some of the interviewees stated that this was a fairly common practice for them. One IBC official observed that principal investigators reporting to his committee often contacted Bethesda for advice preliminary to filing papers, a fact that failed to surface in discussions with researchers themselves. Respondents were unanimously appreciative of ORDA's assistance in processing their interrogations. They also agreed that NIH personnel had never performed on-site inspections of their records, nor had they monitored the experimentation process in any way except to pose an occasional inquiry as to the contents of a research proposal. To the extent that there is guideline implementation, it is done by campus personnel.

When the matter of IBC project supervision came up, however, a wide variety of enforcement strategies, some of them not even enforcement patterns, emerged, just as the comments elicited from cloners forewarned. Panel chairmen divided 4 to 3 on whether their review boards employed any procedures to monitor gene-splicing activities, a minority responding in the affirmative. "Our biosafety officer inspects laboratories," one said, "and sees that the doors are secured and that the logs are kept." But another reported a much more elaborate process: "The biosafety officer examines the physical layout, conducts tests to determine staff expertise, and directs an annual recertification regimen. When I went to Washington for the meeting of IBC heads, I was shocked to learn virtually no other institution provides these checks. It seems to our committee that the guidelines are clear; oversight was never supposed to be left in the scientists' hands. However, our job is not *research* supervision, it is *safety* inspection." Just what implementation efforts does the majority consider appropriate for their IBCs? "Committee members, as they walk through the building socializing, can see obvious infractions. Once, when doors were open, we told a colleague doing P2 work to be more careful." "We don't have any P3 research, so we really don't need a biosafety officer. Still, we made sure one of our people had the right training, and she occasionally looks at P2 facilities." Most committees, then, seem to rely on an informal drop-in approach, many taking the

Table 4.3. Classification of projects as pure or applied research by principal investigator, IBC, and author

	Researcher's evaluation	IBC evaluation	Author's evaluation
Scientist A	pure	pure	pure
Scientist B[a]	pure-pure	pure-pure	pure-pure
Scientist C	pure	pure	pure
Scientist D[a]	pure-applied-applied	pure (package conjecture)	pure-applied-applied
Scientist E	mixed	pure	pure
Scientist F	applied	pure	pure
Scientist G	pure	pure	pure

[a]Multiple projects.

position that the hazards attendant in the cloning projects for which they are responsible do not pose special problems. Naturally, the likelihood that any controversial issues would surface between the IBC and the researcher are few and far between under these conditions; in fact, only one panel representative enumerated such conflict, the question being whether the biosafety officer could require a P3 researcher to dispose of waste materials his work had produced. Review bureau heads mention only a single instance of collegial whistle-blowing, the allegation proving, upon investigation, to be without foundation. So both scientist and administrator agree that conflicts between them have been minimal at most, but whereas committee spokesmen see this as a sign of mutual respect and understanding, many researchers, as I have reported, view the level of oversight as unprofessional, if not irresponsible.

I asked each panel chairman to classify as pure or applied research the experiments of a particular principal investigator-interviewee working under his committee's jurisdiction.[15] It was not the purpose of this request to obtain an expert, independent judgment which might then be contrasted with the views of both researcher and writer, though it could conceivably have accomplished that aim. Rather, I sought insight into whether scientist and institutional guideline enforcer look at recombinant DNA projects from the same theoretical vantage point. Recall that the pure-applied distinction is not only fundamental to conceptualizing the nature of research but also triggers burdens of proof regarding the constitutional standing of scientific endeavors. Table 4.3 gives the classifications accorded the research designs of seven principal investigators

15. Where I interviewed more than one recombinant DNA scientist at a particular institution, the researcher whose work I discussed in depth with IBC officials was the first person drawn at random from that campus's pool of gene splicers. See Appendix A for the procedure through which cloner interviewees were selected.

by the investigators themselves, by the IBC spokesmen, and by this writer. Scientists B and D were involved in multiple projects, so that the IBC respondents were being asked to characterize ten research sets. In seven instances (scientists A, B, C, and G), panel chairmen (and this writer) described the gene-splicing enterprise as basic science, agreeing with the researcher on five occasions. In two cases (scientists E and F), both the IBC official and I disagreed with cloning specialists, perceiving their research as pure whereas the investigator classified the subject matter as either applied or mixed. Given the fact that review board chairmen are trained in the natural sciences and not in constitutional law, I anticipated a greater agreement between administrator and investigator than between political scientist and recombinant DNA researcher. Evidently, the specialization distance separating the insider from the outsider is such that even colleagues trained in the general area of biological phenomena assume the status of well-informed laypersons when they attempt to grasp the central thrust of cloning exercises. One may speculate on whether academic administrative officials realize the extent to which applied recombinant DNA experimentation—if such labors are correctly catalogued by those responsible for developing them—is going on under institutional auspices. Certainly the oversight brought to bear on scientist D's research does not brighten the prognosis. When asked to classify that professor's studies, the chairman whose panel has jurisdiction over them, announced, after rummaging through committee files: "His protocols seem to have been misplaced, but knowing him as I do, I would expect his work to be basic inquiry." It is just this kind of bureaucratic disorientation that some of our gene-splicer interviewees found appalling. And, in this case, the chairman's conclusions were almost certainly dead wrong. All in all, though, the shared attitudes manifested by these ratings augur well for the proper application of free expression doctrine to provocative scientific investigations, whether these be essayed in the courtroom, the classroom, or the bureaucrat's office.

In her review of California biosafety groups, Dutton found that 83 percent of panel officials rated the NIH guidelines "about right," with only 17 percent finding them unduly restrictive.[16] My respondents,

16. *Recombinant DNA Research* 6 (April 1981): 200. In Philip Bereano's opinion, IBC panels have two alternative ways of interpreting their roles under the guidelines: they can either confine themselves to technical questions or branch out and address "the impact and implication of the work [including its] social/political ramifications." He also takes the committees to task for doing so little of the latter. See "Institutional Biosafety Committees," 20. I see no evidence that NIH ever intended its campus surrogates to indulge in a study of sociopolitical issues when passing upon proposed genetic engineering projects. I also submit that it is a clear violation of First Amendment liberty to reject a grant application for basic research of any description, cloning or otherwise, because federal spokesmen

speaking two years later, were very closely divided. A slim majority (4) said the constraints have been appropriate, while two disagreed and one concurred in part and dissented in part. The prevailing view was that deregulation had been sound and that further revisions would be in order as greater knowledge emerged. Three argued, however, that the constraints were and are excessive because demonstrable hazards have never been shown, though one of these bureau chiefs felt that the guidelines are not oppressive because principal investigators can readily comply with them. We have here further evidence that the "debate" over mandatory restrictions has never been livelier, even among policymakers.

All panel chairmen hold their review boards' track records in very high esteem, the only self-criticism as to substantive issues taking the form of minor second-guesses respecting decisions made early in the game. "At the beginning, there was some confusion about our role, but we adjusted well." And there was this comment: "If anything, our members were *too* conscious of community sentiment. We issued press releases and went on the radio. The hazards were overrated, but the mayor's office thought we handled everything just right because we stopped protest movements before they started." On the matter of oversight efficiency, a sprinkling of negative remarks showed some frustration with Bethesda's supervisory strategies. "We have been efficient, but NIH's twelve-step process for approving memoranda of understanding was absurd." However, a large majority had no trouble approving campus regulatory methods. Only one spokesman said that investigators who griped about local standards of efficiency would have just cause. "It takes too long for experimentation to get off the ground. Also, we don't keep current in any systematic way with what cloners are doing, but we can't just nose around." Of course, one would expect any regime to defend itself, and one ought not be surprised when breakdowns in the system are charged to "the people upstairs." Still, the IBC chairmen displayed a curious lack of "feel" for the complexities of researchers' dissatisfaction. "Local control" is simply not a synonym for "shared political wavelength," a lesson which neither Bethesda nor many of its campus surrogates have learned.

We have explored from several vantage points the responsibilities which NIH officials have placed on IBC panels, yet we have not examined the relationship between those panels and the academic institutions they represent. To what extent must these committees tailor their policymaking and enforcement authorities to larger political influences

---

find the study objectionable on sociopolitical grounds. This is content bias pure and simple. Cf. Richard Delgado et al., "Can Science Be Inopportune? Constitutional Validity of Governmental Restrictions on Race-IQ Research," *UCLA Law Review* 31 (1983): 128–225.

and power structures on the campus? If review board spokesmen agree on anything, it is that the groups they lead enjoy a very wide range of discretion in these respects. Only one respondent said that the university administration imposed careful standards of oversight on panel operations, and none reported that other centers of control within the institutional community attempted to influence either bureau deliberations or follow-through. It is noteworthy—and, I think, salutary—that the single case of close supervision involved the work of the campus biohazards panel that is actually a blue-ribbon stand-in for the duly constituted IBC.

This section concludes by ascertaining whether chairmen feel it is time to make the guidelines voluntary, and whether they believe the Asilomar-NIH strategy was a mistake. We then can compare their views to those offered by gene-cloning researchers. Bearing witness once again to their lack of consensus, panel representatives split 4 to 3 in favor of retaining mandatory controls under Bethesda's stewardship. For the minority faction, a sufficient response was "It's time to go all the way." For the majority faction, more elaborate justifications proved necessary: "You need a framework, and national controls are better." "Cloners are like automobile owners; they should be licensed. Safety is the common denominator." "If seat belts should be required of someone driving a car—and they ought to be—then recombinant DNA scientists also should be harnessed by rules." Those who presume that the new wave of scientists is interested only in aggrandizing professional objectives and pushing back the frontiers of research would have considerable difficulty squaring this wafer-thin vote of confidence in the status quo regulatory blueprint with the 74 percent approval figure reported earlier for principal investigators. As a matter of fact, in November 1980, one year before these interviews, campus committee chairmen conducted a straw vote on this issue at the second of their meetings in Washington, D.C., and those participating favored abolition of the obligatory element in the guidelines. On the other hand, our researchers might very well take umbrage at the suggestion that scientific experimentation addressing the nature of biological species is properly analogous to operating a motor vehicle. Safety claims, as we determined in Chapter 2, cannot automatically be accorded irrebuttable status, if the United States Constitution is to protect the right of scholarly inquiry. But even the chairmen who are ready to strip Bethesda of all its regulatory teeth in making the guidelines meaningful reject a similar emasculation of campus authority. "When cloners use hazardous materials, someone has to see that the work is contained properly," was their principal defense. In this determination to keep tight the reins of institutional surveillance, our "liberal" IBC contingent contrasts vividly with our "liberal" researcher bloc, 80 percent of whom want all restrictions and oversight to be strictly voluntary. If self-interest is indeed the prime element in framing attitudes toward modes of public policy intervention, I fail to perceive much difference between the scientist who chafes under outside constraints per se and the local administrative official who chafes under federal hegemony.

But perhaps a better test of self-interest is provided by the question of whether IBC personnel, in concert with the vast majority of cloning investigators, think NIH's grand supervisory strategy has justified itself. Our small sample is prepared to go this far only by a whisker, citing the same sorts of reasons which scientist respondents enumerated, namely, fear of congressional legislation and fear of unknown pathogenic consequences. The minority position was best summed up with this assertion: "Gene splicers made a mistake in not consulting microbiologists who have been studying these organisms for 100 years without legal codes."

The salient conclusion emerging from all this is that recombinant DNA researchers as a whole are much more receptive than are IBC chairmen as a whole to the NIH model of review and oversight through binding norms of laboratory decorum.[17] We cannot infer that cloners are any more public-spirited than their watchdogs, nor can we infer that these administrators have any greater concern for the marketplace of research ideas than do the scientists they supervise. And, as we have seen, frustration with Bethesda's rigors has informed the convictions of both groups; moreover, researchers must put up not only with NIH but with what they sometimes regard as the vagaries of IBC oversight, something they appear perfectly willing to do. At this time, conjecture should perhaps be reserved for the obvious: recombinant DNA workers have been, and are, running a little scared. They fear adverse public reaction; they fear oppressive legislation; they fear local harassment; they fear carelessness by irresponsible peers. And it is upon these anxieties that the Asilomar consensus, as an exercise in enlightened self-interest, was and is predicated.

Clones as Pots of Gold

My commentary toward the close of Chapter 3 emphasized the burgeoning relations between the worlds of campus and commerce, as each seeks to capitalize on the potential for profit of recombinant DNA technological sophistication. But these trends and tendencies cannot be described —much less evaluated—with full clarity merely by setting forth the terms of a contract here and a transaction there. A missing link is the constella-

17. Further proof consists of yet another informal tally recorded at the 1980 convention of IBC chairmen in which an overwhelming majority thought that their panels would be well advised to hand over their responsibilities to health safety inspectors. Milewski, "Report," 25.

tion of relevant attitudes and values subscribed to by members of the academic cloning community. Are they prepared to go partners with industry? Do they want their universities to use gene splicing as a mechanism for enhancing their endowments? Is their research for sale, and at what price to the state of educational and scientific integrity? How do their judgments contrast with those of the NIH-campus administrative hierarchy which would be called upon to adjust the guidelines to these new political arrangements?

By no less than a unanimous tally, the gene splicers to whom I talked thought it appropriate for academic entities to negotiate contracts with private enterprise for subsidizing recombinant DNA endeavors. Two reasons dominate respondents' thinking. One is frustration over the attenuation of NIH funds; again and again, the term "Reaganomics" was bandied about contemptuously. The second appears to smack a bit of professional jealousy, and can be summed up by borrowing from an old Cole Porter lyric: "The chemists do it; the physicists do it; the pharmaceuticals are up to their ears in it; why should we be held to a higher standard?" On this question, the watchdogs with whom investigators work were generally supportive, noting as well the financial crunch which they thought could stifle worthy proposals.

I inquired as to what guidelines practitioners felt should structure these ties, focusing particularly on two dimensions with obvious institutional ramifications—academic freedom and patent rights. Respecting the First Amendment variable, by far the most enumerated preferred values were "freedom of publication" and "freedom to exchange ideas." There has been some talk lately about faculty who present their findings at professional meetings but refuse to divulge important aspects of their investigations on the ground that the information is proprietary, i.e., potentially patentable.[18] Here we have scientists forced to choose between two rights, the first a constitutional prerogative to disseminate ideas, the second a statutory freedom to obtain or sell access to "unique vendibles." My data indicate that in the academic world, selection of the latter over the former, thus clogging the replicative process, would be considered a violation of implicit canons. Other equally fundamental First Amendment aspects mentioned were the freedom to choose one's own subject for analysis and the freedom to direct the course of research exploration. I suspect these were named less frequently than the others simply because they are taken for granted by a vast majority. Those four elements, then, seem to constitute for respondents the guts of scholarly enterprise, and I submit that recombinant DNA academics would not sit still while their university peers bargained them away.

18. Nicholas Wade, "Gold Pipettes Make for Tight Lips," *Science* 212 (June 19, 1981): 1368.

Several interviewees also addressed the dangers of such arrangements from the financial sword of Damocles standpoint. The following remarks bespeak concerned cloners' attitudes: "The MGH-Hoechst contract is acceptable, because 'inquiry' and 'technology transfer' are carefully demarcated." "We should be able to consult but not be corporate executives." "A clean separation between those funded by industry and those funded by NIH is basic. Genentech was established by U. Cal. professors feeding at the taxpayers' trough." The essential fear is that industry will purchase control over the idea markets it so badly needs, and the first line of defense is perceived to be a checks-and-balances calculus. The cutting edge is exemplified by the MIT-Whitehead agreement, which received no votes of confidence and several shrugs of uncertainty. The greatest collegial indignation would ostensibly be reserved for the scientist who did applied, on-campus research and received concurrent support from the taxpayer in the form of a grant, from industry in the form of stock options, and from the university in the form of salary increases made possible, in part, because the university-employer owned an interest in the researcher's lucrative investigations.

Interestingly, panel chairmen seem to have thought these matters through somewhat more carefully than have practitioners, and perhaps as a result, they insisted on more rigorous safeguards for academic interests. Of course, there were the expected warnings about contracts that stifle scholarly communication and profit-oriented professors whose primary work is nonacademic. But among these respondents the MIT-Whitehead agreement was denounced unequivocally, while the sole comment about the Hoechst transaction was also highly negative. Another spokesman concentrated his fire on Herbert Boyer and Walter Gilbert, calling their relations, while on university appointment, with Genentech and Biogen, respectively, totally unacceptable. Finally, the argument was advanced that university arrangements with industry respecting recombinant DNA work should not be fashioned unless the academic departments involved participate in the rule-making process. To generalize, the consensus among IBC chairmen was that the private sector has an interest only in the dollar bill and that, absent a recession in the grant market, universities are ill-advised to deal with them; however, as things stand, academia will gain more than it loses by taking a fee for the enhancement of technology transfer, but only under carefully circumscribed standards.

As could be expected, the cloning community had considerably more difficulty engaging the question of appropriate standards of practice for determining patent rights. Only 47 percent could propound specific policies, and a mere 32 percent advanced highly organized and articulate notions. The dominant opinion—that the academic institution should file for patents while private outfits could obtain licenses, develop products, and pay royalties—is certainly well grounded in experience, but basically the legalisms involved are as yet too arcane for a community of scholars who have suddenly discovered incentives to hone their entrepreneurial skills.

I thought IBC officials, because of their administrative experience, would negotiate this terrain with greater dexterity; however, this proved not to be the case. Half of them firmly believe that academic institutions must hold whatever patent claims develop and that companies should be able to dicker for licensing rights. But regarding appropriate rewards for the principal investigator under these arrangements even the most thoughtful and knowledgeable bureaucrats encountered difficulties. On the one hand, to promise scientists royalties would certainly encourage them not to communicate with their peers; on the other hand, to give universities, but not investigators, financial bounties might well prompt quality people to leave higher education.

This matter of research practitioner payoff does not seem to bother research practitioners very much, though. Above, we saw that there was no great hue and cry on their parts extolling the notion of patent rights as an attribute of scientific discovery. That conviction received further amplification when I sought to ascertain whether cloners considered their contributions to be their own private property or part of the idea marketplace. Of the twenty-three studies under development, scientists rated 83 percent as involving the acquisition of knowledge that should be accessible to all. Some of the commentary is edifying: "The NIH is paying for my work, not industry." "If I found something patentable, I'd quit and start my own company. That's what Arnold Beckman did. But you can't have your cake and eat it, too." Recall that researchers characterized seventeen of their twenty-three projects as exercises in basic experimentation. With respect to sixteen of these, plus three designs that were rated either "mixed" or "applied," scientists said that the objectives and accomplishments they hoped to achieve belonged in the idea marketplace. Our cloners, then, see themselves as educators doing pure research and increasing the common fund of understanding. If universities can negotiate financially lucrative arrangements with commercial enterprises, fine; but I think it unlikely that these scientists would suddenly reincarnate as company employees, churning out patentable vendibles for a profit. Even the small minority had its share of reservations. "My work is applied research, but I think any money I make from patents should be recycled to academic departments. I am a professor, after all." "I do basic science and I publish, but I've produced inventions, too, almost by accident. The university has gotten patents on these, which is certainly appropriate. So my end-results are an 'idea-property' mix." If compacts between industry and higher education are a new focus of recombinant DNA debate, then I can at least report that there is a maximum degree of leverage for compromising divergent views of patent privilege but that cloners are no more ready than anyone else to sell the academic store.

A closely related policy issue is whether universities should permit their researchers to develop personal funding agreements with private corporations in order to merchandise whatever wares they contrive. I expected my scientist respondents to be more critical of this laissez faire model than they were of the *parens patriae* strategy because, at least in theory, it presents a greater likelihood that the community of scholars ideal will be sacrificed on the altar of pluralistic enrichment. The predicted drift is discernible, but slight. Institutional endorsement of side-payment bargains win the approval of about 90 percent of cloners, most of whom either find these liaisons supported by ample precedent or simply say, "My outside work will help keep me in academic life." There was recognition that such ties might turn some laboratories into islands, but the thought seemed a trifle abstract to most, for, as one commentator noted, "As things stand now, we usually don't talk until we're good and ready." Not lacking in reality, however, was the Walter Gilbert case; a few interviewees mentioned his situation, and he came away with only minority support. Gilbert, a Nobel Prize molecular biologist, had wanted to retain his professorship at Harvard while also serving on the board of directors for Biogen. At the time these interviews were conducted, he was on well-publicized leave of absence from university duties, evidently to rethink and renegotiate his future professional alternatives. "Gilbert is no longer an academician," one respondent opined; "he has become another Kissinger." But to reiterate: The vast majority would be very pleased if business interests offered to subsidize their work and take as a reward the option of marketing whatever useful artifacts and technologies arose therefrom. Nor, apparently, would they reject either a little stock in the company, if commercial prospects loomed large, or "kitchen cabinet" status in the conference meeting room. How many would go further and opt for de jure brain-trustership in the world of venture capitalism, betting on their own reputations and the careers of others, I cannot say. But if this is essentially the vice of Gilbertism, it would be incorrect to argue that a consensus among recombinant DNA scientists accepts it uncritically.

It is on this item that IBC chairmen, perhaps because of their quasi-

Table 4.4. Cloners' preferences as to funding sources

Rank	Funding source	Weighted score[a]	Score if ranking were unanimous[b]
1	Federal government	26.5	(19)
2	In-house royalties	40.5	(38)
3	Private industry	47.0	(57)

[a] First-place selection = 1 point.
Second-place selection = 2 points.
Third-place selection = 3 points.
[b] Number of points if all cloners placed option in this position.

company perspective, begin jumping off the researchers' bandwagon. Their sentiments were evenly divided, the feeling that "academics are not supposed to be entrepreneurs" being balanced against the belief that engineers and chemists cannot have special privileges, even though their efforts are often much more in the applied sphere.

I also sought insight into principal investigators' personal philosophies regarding money sources. I assumed that the typical subject was committed, all else being equal, to the professorial life, and I think the data bear this out. Unfortunately perhaps, all things are not equal in the world of recombinant DNA research support. Faculty gene-splicers would like nothing better than full funding from the home institution, but there is not a school in the country that can give much more than a handful of experimental scientists the financial backing they need out of general revenues. Enter NIH with its red tape, guidelines, and peer review; enter business with its nose to the ground in search of product. I asked cloners, assuming they had their druthers, how they would rank-order their preferences from among three funding sources: the corporation, the federal government, or campus monies derived from royalties accruing through patents which colleague genetic engineers had developed. This last option has been called the Wisconsin plan, because biochemists at Madison have built several first-rate departments using as their chief financial buttress resources in excess of $100 million earned as a result of patents held on the process to isolate Vitamin D and on the anticoagulant Warfarin.[19] Lately, it has received some play as the possible format under which recombinant DNA science could salvage both its academic dignity and its financial independence. Awarding one point for

19. For a description of the Wisconsin plan, see E. B. Fred, *The Role of the Wisconsin Alumni Research Foundation in the Support of Research at the University of Wisconsin* (Madison: WARF, 1973), and the 1975 supplement to this report by William Jordan.

a first-place selection, two points for a second-place pick, and three points for a third-place designation, the totals are as shown in Table 4.4. Cloners accorded federal government subsidies a healthy fourteen-point margin, including twelve of nineteen (63%) first-place tallies. The Wisconsin formula, predictably, did better than the corporate sector, but could only be said to have finished a disappointing runner-up, and some might be surprised that industry did as well as it did. In first-place choices, the in-house royalty model collected only two votes to private enterprise's one, so the difference can be found almost entirely in the second-preference column, which the Wisconsin approach garnered by a count of twelve to six. Of some interest is the fact that one-fifth of the sample failed to select an optimal preference, and in a few cases the response was, "I don't care; I'll take the money from the people who give me the most freedom. All of them can make life difficult." Given the ups and downs of the recombinant DNA dialogue, such pragmatism (cynicism?) among a minority can neither be considered unanticipated or unhealthy.

Why did respondents evaluate these options as they did? Federal support, which for our scientists means NIH support, has, they said, a proven track record, emphasizes rewards for basic research, and evaluates proposals in an objective and professional way. There were a few grumbles about "paperwork," the "buddy system," and declining project opportunities, but positive remarks outnumbered the negatives by at least two to one. The Wisconsin option suffers, in their minds, from several defects. The most prevalent opinion was that it is simply unrealistic; as bastions of pure science, university communities could never raise sufficient funds from patenting recombinant artifacts to cover necessary costs and needs. Others thought that decisions on how to allocate monies among competing specialists would cause endless departmental bickering. Its attractiveness when weighed against the option of corporate support stems from the preference of our sample to resolve all doubts in favor of the academic environment. Business, of course, offers the most enticing pecuniary rewards, and cloners agree it places fewer strings on an investigator's autonomy. But the string of immediate technological payoff and the lack of security for those who fail to produce make industry a not very palatable choice for these academics, unless they can pretty well write their own tickets.

We have seen that federal authorities, for all their concern about cloning in the academic environment, have not extended mandatory guidelines to industrial gene-splicing. I queried the panel of researchers on this anomaly, and found that 74 percent thought such government inaction constituted a loophole in need of plugging. Actually, given the lack of even-handedness in federal policy, I anticipated that principal investigators would express unanimous displeasure on this point. But four of the five dissenters constitute 80 percent of the group that advocates repeal of the national mandatory guideline component for campus work. Their attitude, then, is one of contempt for these constraints as a general proposition, because they fail to discern any danger to the public no matter where the research is conducted. Of the fourteen who favor guideline extension, a majority (63%) believe the NIH restrictions would suffice for profit-making enterprises. "Hazards do not rise or fall with laboratory ownership," ran the typical argument endorsing across-the-board supervision.

Certainly IBC heads would have no reason to exempt Monsanto and Du Pont from federal regulations, and I thought they would concur pretty much in the researchers' sentiment. The data indeed show an almost identical response, as 71 percent approved industrial guideline coverage. This majority also tendered unanimous endorsement of coextensive rules. The notion that academic environments ought to receive a benefit of the doubt not accorded commercial enterprises in the regulatory arena receives only token recognition with chairmen as with scientists.

Recombinant DNA investigators sit in a unique position, when judged against the backdrop of the history of science. Their talents can result almost overnight in both intellectual breakthrough and material affluence. If they labor in a house of learning, though, their scholarly ambitions may be thwarted by the hard realities of economy. If, on the other hand, they cast their lot with industrial firepower, they may well have no scholarly life at all. Our geneticists display a healthy regard for, yet an enlightened suspicion of, what the world of commerce can provide. They think industrial-university diplomatic relations in this field are unfolding at much too leisurely a pace either for the good of their discipline or for the good of academic solvency. They know bargains must be struck to protect the basics of the educational enterprise, because they want to remain a part of that enterprise. They have a coherent sense of what academic freedom means, and it is a sense that is largely shared by those administrative actors most familiar with what they do. Thus, two likely prognoses regarding the financial state of cloning research as pure science are in order. One is exemplified by the headline transaction, MIT-Whitehead being the prime example. The other is the less compromising model envisioned by a majority of academicians, whether they be specialists or policymakers. Fortunately, prognosis in this context is a function of attitude, value, and choice. As the analysis in Chapter 3 made manifest, I share their view that the latter prognosis stands not only as a more desirable but also as an achievable outcome.

The Constitution in the Laboratory:
An Empirical Analysis

The Constitution, I have argued, is more than a scrap of parchment; it is a "living" document. Under the American scheme of republican governance, the people can change the rules of the political game by formal amendment, should they so choose, and the courts can update the meanings of constitutional language in the course of deciding who wins legal controversies. Moreover, customs and usages arise, sometimes without the citizenry being the wiser, taking on wills and artifices of their own. A constitution may also be seen as a way of looking at political life, at political purposes, at political responsibilities. In this sense, constitutionalism is akin to theory, but in the context of my analysis, I have downplayed the philosophical element and emphasized the social science implications. The Constitution, I have assumed, connotes valued principle, notions of right, obligation, and procedural nicety which public opinion places above the usual run of political activity and accommodation.

The concern here, however, is not with what the public thinks, but with what scientists think, and a particular breed of scientist at that. The literature informs us that scientists have become a political interest, following in the footsteps of farmers, labor, management, and other vocational groupings.[20] Unfortunately, that literature conveys little information about the constitutional-value universes in which these people move intellectually and, perhaps, psychologically.[21] How do members of particular factions perceive the Constitution as a set of norms ordering their behavior and the behavior of salient political actors? That question becomes especially important in this study because the scientists whose work ways we are exploring must deal with the state, its personnel, and its routines as a matter of course, else their research will likely fall by the financial wayside. We would not expect cloners, any more that we would expect their watchdogs, to be overly concerned with constitutional proprieties as a specific set of values against which they monitor their latest political attitudes and actions. But we would expect both groups to acknowledge and be competent to articulate basic standards of right con-

20. Don K. Price, *The Scientific Estate* (Cambridge: Harvard University Press, 1965).
21. For basic political values as psychological variables, see Robert E. Lane, *Political Ideology* (New York: Free Press, 1962).

duct which color their perception of scientific activity in general and of recombinant DNA endeavors in particular as socially relevant phenomena. To the extent that these norms are seen to order significant relationships between science and government, they are, as I have repeatedly contended, living constitutional material. Let us see if we can isolate their parameters.

The first question we address could hardly be more important: do interviewees perceive scientific investigation as a First Amendment value? A significant majority among researchers (63%) agreed that it was, while 21 percent disagreed and 16 percent said they didn't know. The standard reason volunteered in support of this view roughly approximates my argument in Chapter 2, viz., the acquisition of knowledge is a fundamental freedom which cannot be curtailed unless special circumstances obtain.[22] The negative responses, though few, were enlightening, because the typical counterargument—"There is no constitutional right to make neutron bombs"—is framed in the conviction that the First Amendment is an absolute. That erroneous assumption, of course, dictates the answer tendered, once we stipulate—as all would stipulate—that some forms of science are subject to restraint. Only one principal investigator who addressed directly the thesis presented in Chapter 2 rejected it, concluding that experiments in the laboratory are conduct, not expression. IBC chairmen were even more supportive of the proposition that scientific inquiry is free expression, with 86 percent registering endorsement. This group demonstrated a much better notion than did respondent geneticists of the First Amendment "balancing" process that courts today find so demanding but so necessary, as the following comments demonstrate: "Scientists have constitutional protection, but they can't do research that poses immediate threats to the community." "Government can regulate physical hazards but has no power to tell William Shockley that his research is out of bounds." Clearly, a consensus among those who are intimately involved with recombinant DNA work on the campus, whether researchers or overseers of research, discern nothing untoward or bizarre about the nexus between scientific liberty and constitutional liberty.

I expected recombinant DNA principal investigators to rate their own projects as a First Amendment value unless they specifically rejected the

22. Three scientists asked for a rundown on what freedoms the First Amendment protected and, when told by this writer "free expression," offered affirmative replies. I do not feel this prejudices the results, for in interviews as in life, people rationally may reserve the option of asking intelligent questions before they formulate their opinions. In retrospect, however, I would rephrase question 29 to read as follows: "There has been some debate lately about whether scientific research is a form of free expression protected by the United States Constitution. Do you think scientific inquiry is entitled to such protection?"

notion of science as a First Amendment value.[23] That hypothesis proved accurate. A typical response was, "We are protected because we pose no danger to anyone else. I don't work with human subjects, and I am not an atomic scientist." Panel chairmen, on the other hand, might conceivably have been a bit leery of characterizing the studies they supervised as free expression, no matter what they thought of science in general. Such rethinking failed to materialize, however. A healthy majority argued simply that the investigations under their administration posed no recognizable hazard; hence, they were protected.[24] There is evidence, to be sure, that a small minority of these projects lack First Amendment coverage because their basic thrust does not constitute a quest for truth. Recall, though, that IBC personnel were unaware of applied research being conducted under their roofs. It is also noteworthy that those involved in technologically oriented studies were just as likely as those involved in basic science to perceive their efforts as free expression. The crucial point, then, is that cloning guidelines are implemented on the campus within a context of shared constitutional beliefs respecting the relevance of First Amendment theory, both scientists and administrators well appreciating the general applicability of that frame of reference.

In Chapter 2, I developed a notion of constitutional privacy which embraces at least some research activities in the home. When queried as to whether they thought the laboratory environment implicates privacy interests of constitutional proportion, gene splicers split almost down the middle, with 47 percent responding in the affirmative and 42 percent replying in the negative. Significantly, members of both contingents almost invariably addressed the proposition from the standpoint of their own work, that is, investigations performed under conditions necessary to contemporary experimental research. Scholarship as a corollary of personal enrichment—comparable to such other "private" behaviors as reading and family planning—is simply unknown to our group and, I suspect, to scientists generally in this day and age of professionalism. Naturally, then, the constitutional concept of privacy means something rather different to the majority subsample than it does to the Supreme

23. One would not expect anyone to argue that science lacked free expression relevance, but that their particular cloning exercises merited such classification. To make sure, however, I followed up two equivocally framed negative answers to question 29 by posing question 30. In both instances, replies were "no." On the other hand, researchers might say "don't know" or "can't comment" to the former, more theoretically oriented item, while responding with great conviction to the constitutional posture of specific studies, especially their own.

24. Two interviewees answered "don't know" to this query, but a panel chairman who couldn't reply to item 34 replied affirmatively here.

Court. For this group, the essence of privacy is that only duly constituted authority can leaf through their experimental plans or monitor their work. A chief villain is the Freedom of Information Act, which they think permits the media, public interest groups, and just plain curiosity-seekers to treat their laboratories as though they were public forums simply because they have accepted taxpayer largesse. "Our studies do not hit the street until we publish," was a common theme sounded by the slender majority. To the minority, though, university research facilities are indeed open to outsiders, especially given the prevalent mode of public subsidy. All cloners who thought the First Amendment had no relevance to doing science also believed that privacy interests had no bearing on doing science. And once more the strain of absolutism emerges; how, some pondered, could any right of privacy exist when researchers can be forbidden to tamper with dangerous substances? IBC personnel were not so badly divided on the issue, but in this instance those leery of privacy protection held the advantage, 57 percent to 29 percent. Again, however, our watchdog sample came closer to the heart of the matter as it is commonly understood. Said one member: "A laboratory is not a bedroom." And another opined: "Science has no bearing on one's personal, private life." However, a dissenter offered the following reflective comment: "Unpublished scientific research is much like unpublished poetry. I see experimentation as an art form. That is how science used to be and should be. Unfortunately, it has become big business."

To most objective analysts, judicial interpretation of privacy rights constitutes one of the murkier areas in American constitutional law. A notion of privacy interest as fundamental norm is not foreign to our respondents, but there is certainly no consensus as to what the concept encompasses. The most one can say is that the front-line practitioner has a very different feeling for salient privacy questions than does the administrative overseer, who turns out to be much more alert to what the courts think are the compelling privacy questions than does the scholar of genetic research. To generalize more broadly, the concept appears still to lie in an inchoate stage, part of the Living Constitution as a vague sense of good form but, even to alert publics, not yet a firm set of understandings.

From the previous discussion of the First Amendment question, we would naturally anticipate all researchers who thought the constitutional right of privacy applicable to laboratory work in general would find it applicable to their own experiments, and in the case of committee personnel, to the cloning done under their jurisdiction. That proved to be the fact, without exception. But some of the remarks tendered are again revealing. On the matter of the right to know, one chairman reflected: "Specific details of ongoing research are shielded under privacy rules. The public has a limited right of access *through our committee.*" As I have indicated, this is a representational model with which the typical gene splicer seemingly can live. And note the following swipe at a "higher law" theory of laboratory confidentiality: "Scientists who make a fetish of privacy don't make the big breakthroughs. I welcome people to share their ideas with me and I'll reciprocate. We live in a market-place; let the best man win!" This last bow to scientific research as athletic competition notwithstanding, I find it hard to accept the belief that privacy and free expression are somehow in irreconcilable conflict.

The next step is to diagram the manner in which these attitudes toward cloning as a constitutionally relevant behavior translate into attitudes toward specific legal constraints put upon scholarly research. I asked these scientists to assume they were funding their own experiments, and then I ascertained to what extent they thought the state could regulate their activities. My question was framed deliberately to exclude consideration of any controls predicated on government's string-attachment prerogatives. The replies collapse into four categories: (1) government has no power to enact regulatory standards; (2) government can pass measures to protect health and safety, but must first demonstrate the actual existence of hazard; (3) government has discretion to enact constraints based solely on rational probability of hazard; (4) government has virtual carte blanche: the academic world stands on no higher ground than the business world when the issue is freedom of experimentation. The numerical breakdown finds the polar constitutional positions (categories 1 and 4) receiving only 32 percent of the vote; but if we array those who presume in favor of the freedom to clone (1 and 2) against those who presume in favor of state action (3 and 4), the margin is 53 percent for the former and 47 percent for the latter, or as close as the numbers allow, assuming full participation. We may surmise that the consensus hospitable to recombinant DNA investigations as constitutionally protected phenomena has not, as yet, been converted into a viable systematic expectation respecting statutory standards. Fluidity and watchful waiting are the bywords.

The question then comes whether our gene splicers are consistent in their views, that to know their general predisposition toward constitutional aspects will allow us to predict their responses to specific legislative frames of reference. If we compare researchers' opinions respecting free expression and privacy coverage of the experiments they are doing with their standing in the above fourfold classification scheme, isolating how many "yes" responses to these two queries show up in each of the first

Table 4.5. Correlation between cloners' views on state regulation of gene splicing and on gene splicing as deserving free expression and privacy protection (number of replies)

	Responses on state regulation	Corresponding responses on	
		Free expression	Privacy
Government can't regulate	3	2	1
Government must show actual risk	7	6	5
Government must show reasonable probability of risk	6	4	3
Government can regulate at its discretion	3	3	3
Total	19	15 (79%)	12 (63%)

NOTE: Respondents' views on government regulation were based on the assumption that they were funding their own experiments.

three sets and how many "no" responses show up in the fourth set, the hierarchy outlined in Table 4.5 emerges.

Keep in mind that the NIH guidelines are based on rational administrative deductions regarding the existence of hazard. Recall as well that these guidelines, should they take the form of *penal restrictions,* would most likely pass constitutional muster even were the courts to hold that cloning as search for nature's secrets is a protected liberty. That is because the NIH standards are largely "times, places, and manner" controls on activities employed as ways of knowing. Essentially, then, scientists who support the "probability of danger" test are among those who believe the NIH commands can be translated into city ordinances and state enactments consistent with the fundamental law. In fact, though the analogy is not perfect, researchers who advocate the "actual risk" test are in the mold of Justice Brennan and his compelling state interest doctrine, while researchers who advocate the "reasoned eye" approach are in the tradition of Justice Frankfurter and his balance of interests doctrine.[25] But most importantly, both these tests presuppose a weighing of social need against constitutionally protected civil liberty.

If these considerations be granted, then the following general comments about our data seem warranted. First, scientists' opinions respecting recombinant DNA's possible status as free expression are very good

25. For Brennan's eclectic use of "compelling state interest," see Sherbert v. Verner, 374 U.S. 398 (1963), Shapiro v. Thompson, 394 U.S. 618 (1969), and Cousins v. Wigoda, 319 U.S. 477 (1975). For Frankfurter's "balancing" orientation, see Dennis v. United States, 341 U.S. 494, 517 (1951) (concurring in the judgment).

barometers for gauging appropriate levels of legislative discretion from the constitutional law standpoint. Investigators who think their cloning research is quasi speech tend to insist that the state meet a higher burden of proof than usual before it asserts police power authority. And researchers who think their gene-splicing exercises are just another form of human behavior with social consequences will tend to believe the state has no presumptions whatever to rebut. Second, while opinions regarding recombinant DNA's possible standing as privacy-related phenomenon are good barometers for anticipating favored standards of legislative scrutiny from the constitutional law perspective, such notions tend to lag behind First Amendment attitudes as predictive tools, whether the various stages of statutory rigor are categorized as above or aggregated. Though for these purposes my numbers are small, I think it clear that scientists possess coherent norm structures sufficient to link preferred constitutional theory with preferred constitutional practice, at least at the level of criminal strictures.

Among the IBC chairmen, replies were much less diffused. None argued that government officials could frame laws treating recombinant DNA research they themselves might undertake as though it were being performed for profit, but none talked about the immunity of the laboratory from regulation either. A strong majority would allow the state enough leeway to codify the NIH guidelines, and the others endorsed the validity of rules designed to protect society and workers but opposed restrictions geared to protect principal investigators from themselves. Again, the larger significance of these data is that campus watchdogs strongly support the concept of university-housed cloning experimentation as expression, and that they construct constitutional models of legal oversight reflective of and consistent with this broad attitude.

Given their understanding of prevailing constitutional limitations, we now examine the extent to which researchers believe government can attach strings or constraints to the subsidized cloning studies they are in fact executing. Principal investigators rejected by a tally of 58 percent to 37 percent the idea that the state possesses authority to place any conditions it deems necessary and proper on their experiments. How does one characterize those who would give government funding agencies a blank constitutional check? These scientists think their work merits First Amendment coverage; they even feel their studies merit privacy considerations; and they would not give the state as much power to pass criminal legislation controlling academic cloning as they think could be mustered to control commercial cloning. In fact, they are just as likely to think their laboratories are exempt from such rules as they are to think the state has carte blanche to punish what it labels excessive behavior.

The stand taken by the minority was simply, "It's the government's money; public agencies can do what they please with it." Shades of O. W. Holmes's theory respecting the state's prerogative to fire policemen who give political speeches.

What conditions do majority spokesmen think government cannot affix to their research? Leading the way were various aspects of First Amendment liberty; in fact, taken together, they constitute more than 60 percent of the taboo requirements enumerated. These can readily be subdivided into four more specific sets of objectionable interferences, which I list in order of times mentioned: (1) controls on the right of experimentation—telling investigators when to do this test and when not to do that test; dictating the use of particular sequencing techniques and restriction enzymes; choking off the acquisition of knowledge; (2) controls on where and when to publish; (3) controls undermining political independence and convictions; and (4) controls restraining the exchange of ideas —clamping security strictures on research. The balance of unwarrantable restrictions lay in the due process and privacy areas. The former (25%) included references to rules considered arbitrary, unreasonable, or capricious. Examples cited were summary government suspension of funded experimentation and "nonsense" guidelines, such as instructing recipients to clone only in pink test tubes. Privacy considerations (13%) focused on requirements authorizing government monitoring of inquiry, that is, "snooping" or "looking over one's shoulder" in the laboratory on a continuing basis. Employing the theories and modes of analysis offered in Chapter 2, I concur in the consensus determination that these various conditions structuring the use of taxpayer largesse, as described to me, are indeed unconstitutional limitations on the right to perform recombinant DNA basic investigations.

A principal understanding upon which many of these opinions are founded is that when the government awards grants for pure research as the culmination of a rigorous peer review process, the money tendered must be used in fashions consistent with the givens of scientific inquiry. Gene splicers endorse the conventional wisdom that grants are radically different from contracts, where NIH, say, can delineate goals, spell out procedures, and reserve all manner of prerogatives. To them, this is the world of applied research funding, experimentation which is prima facie nonacademic. But the entire purpose of peer criticism, in their eyes, is to select the most competent, insightful, even creative proposals for looking into the nature of genetic materials and processes. The expectation— indeed, the *constitutional duty*—arising from such a grant is that the people's representatives will permit cloning to be performed scientifically, will insist, in short, that the marketplace of analysis not be shorn of the detachment and robust debate so necessary for proper testing and weighing of facts and theory.

I also examined the argument presented in Chapter 2 that government lacked the constitutional discretion to subsidize genetic investigations while declining to subsidize all manner of recombinant DNA investigations. I found that 63 percent of researchers agreed with this assessment. Some dismissed the proposed restriction cavalierly, bandying about such adjectives as "ridiculous," "inconceivable," "nonsense," all synonyms for unreasonable and, in this context, unconstitutional. But others noted that the two modes of research have become inseparable. Discovery through cloning has achieved so much, they felt, that to restrict financing to what were fifteen years ago considered the conventional "expressive activities" would rip the heart out of the discipline. That is (if I may invoke the rhetoric of constitutional parlance), it would violate both the principle of rationality and the principle of content-neutrality to confine money grants to those physicists who employed only the technologies and instruments of proof known to Newton and his predecessors. Yet, a substantial bloc dissented, surprisingly so given the obvious conflict-of-interest temptations to do otherwise. The locus of opposition came from several gene splicers who had earlier said government retained carte blanche in the string-attachment game: any such exception would be ill-advised, they thought, perhaps even foolhardy, but public officials, should they go off the deep end, were susceptible to lobbying and pressures through the ballot box. The Frankfurterian spirit lives on, even in the laboratory! But almost as many investigators who thought the state had a free hand in placing conditions on money gifts balked when the string at issue would put *them* out of the cloning business. Of course, it would be perfectly plausible for someone to argue that government is limited in contriving exceptions but that a ban on recombinant DNA work is within the scope of constitutional discretion. However, less than 30 percent of scientists who think some strings impermissible take this view.

By a majority healthier even than that posted among our scientists (71% to 58%), IBC representatives believe the state is limited in its capacity to impose strings. Moreover, their recitation of the conditions which government is forbidden to apply reads like a carbon copy of those listed earlier, complete with distinctions between grants and contracts and between the "public hazard" context, where reasonable guidelines are beyond constitutional cavil, and the "intellectual" context, where guidelines of any sort are at once suspect. Nor do the parallels cease when we inquire about government's authority to fund genetics in general while not funding cloning in particular. Just as our researchers rejected this option by virtually the same margin that they rejected the constitutional theory of "unconditional strings"—63 percent to 58 percent—so our administrators opposed both by strikingly similar, and higher, margins—71 percent in each instance. And, not only were the reasons tendered on all fours with what scientists offered respecting the flaws inherent in an anticloning exception, but also the very same kind of switch effect occurred when I compared reactions to this specific restriction with reactions to formulation of restrictions as a general proposition. Hence, 50 percent of chairmen who thought all conditions on subsidy were constitutional thought that this condition was unjustifiable, while 80 percent of those who thought government lacked carte blanche in selection of conditions thought this condition was impermissible.[26] One may say with rather more confidence than seemed possible prior to this empirical analysis that NIH campus administrators share with recombinant DNA principal investigators a rationally structured set of constitutional expectations regarding governmental authority to manipulate its financial levers of support for the purpose of working social and political policy in the laboratory. Even threads of inconsistency cut across interest lines. Of course, there are deep divisions respecting both the propriety of the NIH guidelines and the broad question of how one defines hazard. But the recombinant DNA debate, as a contrapuntal exercise engaging cloner and watchdog, has apparently never exceeded those dimensions.

That these reflections do not suffer from overbreadth is demonstrated by comparing not only interviewees' responses to questions of constitutional theory but also to questions of constitutional practice. All of the IBC chairmen and 91 percent of the scientists said government had neither imposed unconstitutional strings[27] nor ignored due process considerations in restricting the gene-splicing experimentation for which they are responsible. As to whether government agencies have at times violated the constitutional rights of scientists to perform recombinant DNA research elsewhere, again all of the committee personnel voted in the negative, while the scientists reported a drop in constitutional endorsement from 91 percent to 73 percent. In their discussions of specific instances, the names of Ian Kennedy, Charles Thomas, and Martin Cline

26. The numbers here are exceedingly small, but as illustrative of the general trend they cannot be ignored.

27. Recall that in reply to question 9, a clear majority of practitioners (73%) said the NIH guidelines, as applied to their investigations, had been reasonable. It is noteworthy that the percentage of those alleging federal norm violation shrinks by 18% when the term "unconstitutional" is substituted for the term "unreasonable." I think this bears witness to a natural reluctance on the part of the average person, including the average, preoccupied scientist-researcher, to accuse duly constituted authority of violating the sacred parchment.

cropped up on numerous occasions, and not one word of support was raised in defense of any of them. The feeling was unanimous that each knew the rules, had broken those rules, and had been treated fairly. Why, then, do some principal investigators think government has, at one time or another, transgressed the standards of right conduct? Two reasons were offered. In the first place, objection was made to the original prohibitions. It was palpably unfair, the argument ran, that NIH would not allow *any* work to be done, for example, with tumor viruses. Secondly, the machinations of the Cambridge City Council were noted; restraining Harvard's initiative to construct a P3 facility was branded constitutionally defective by those who mentioned that episode.[28]

To repeat: Cloners and their campus watchdogs, sharing a great interest in the well-being of science as a method and as a body of knowledge, appear to have a common sense of what is meant by an unconstitutional impingement upon recombinant DNA investigations. And they also share the view that the federal government, as chief overseer and guideline-setter, has been faithful to these constitutional understandings.

A Footnote in Political Demography

No behavioral investigation of the relationship between cloning and the public welfare would be complete without introducing respondents as educated, thoughtful citizens, holding very definite opinions regarding their place in the American political scheme of things. Chapter 4 concludes by providing this introduction, sketching attitudinal profiles of our two interest groups and noting relationships, if any, among personal attributes and convictions, on the one hand, and the constitutional politics of recombinant DNA experimentation, on the other hand.

A staple item on the political science survey research questionnaire is party identification. More than any other single variable, partisan affiliation is considered a key indicator of voting[29] and issue orientation[30] in the United States. The literature also tells us that college professors, except those working in "business-related applied fields," tend to register strong support for Democratic presidential nominees.[31] Perhaps, then, we ought to be taken aback when we learn that only 32 percent of our scien-

28. I am talking here of only a handful of negative votes, though I suspect that many of the respondents incorrectly interpreted this question as a check on their assessments of federal action alone.

29. See generally Angus Campbell, Philip E. Converse, Warren E. Miller, and Donald E. Stokes, *The American Voter* (New York: Wiley, 1960).

30. Everett C. Ladd, Jr., and Charles D. Hadley, *Political Parties and Political Issues* (Beverly Hills: Sage, 1973), 18.

31. Everett C. Ladd, Jr., and Seymour Martin Lipset, *The Divided Academy* (New York: McGraw-Hill, 1975), 57, 64, 159.

tists prefer the party of Franklin D. Roosevelt and John F. Kennedy. And how can we explain the fact that 63 percent identify with neither major party, 50 percent calling themselves independents and 13 percent calling themselves "other"? The last of these designations presents little problem, as it turns out, because a small minority in our group are resident aliens who confess to feeling no particular kinship toward the American party system. But with respect to self-styled independents, the conventional wisdom codes them as somewhat alienated from the system, their lack of partisanship constituting a spin-off from their lack of involvement generally.[32] These are people, it has been said, who simply "are not much interested in politics and government."[33] Needless to say, one would not anticipate this species of nonpartisan to be long on formal education. And yet there is a new kind of independent as well: very knowledgeable, highly educated, civic-minded, younger people, who have turned away from the two-party struggle because of Vietnam and Watergate.[34] Do 50 percent of our cloners fit this mold? The answer is no. While I lack data concerning either respondents' political issue positions or age, I do have information regarding their academic rank. If the "new independent" hypothesis were valid, one would anticipate not only a positive correlation between full professor status and party membership but also a positive relationship between junior professor status and independent standing. Yet, half of our senior faculty are nonpartisans, and more than a third of our independents are senior faculty,[35]—hardly the makings of statistical significance even though 70 percent of the associate professor-and-under group do indeed consider themselves unaffiliated. Interestingly, when I asked panel chairmen to state their preferences, five of six full professors (and the single associate professor) called themselves Democrats, an evident reassertion of form.

I also sought to determine if there were any connection between interviewees' choice of political party identification and the work they did. An overwhelming proportion among scientists (85%) said no such nexus existed. However, the lone citizen who classified himself as an "other" remarked, "My research is dictated by political consciousness. I'm primarily concerned with developing cheaper agricultural techniques to help poor people." And then there were two independents who described their nonpartisanship in these rather different ways: "The major parties

32. William H. Flanigan, *Political Behavior of the American Electorate,* 2nd ed. (Boston: Allyn and Bacon, 1972), 47.

33. Campbell et al., *The American Voter,* 143.

34. See generally Gerald Pomper, *Voters' Choice* (New York: Dodd, Mead, 1975).

35. Our universe of eighteen excludes, for obvious reasons, citizens of other countries but includes (see Appendix A) two scientists working on the same project.

are irrelevant to science. One panders to the voters, while the other panders to money." "I've never thought much about the parties. Heck, I've never even voted." This last comment came from a senior professor, while the two previously cited quotations were offered by younger faculty. Can it be that my original hypothesis was incorrectly framed, that some senior cloners are so wrapped up in their research that partisan politics correlates not at all with their intellectual mindsets, that many of them are, in fact, old-style independents? More spadework is needed before we confront directly this distinct possibility.

With respect to a possible relationship between the Democratic party predilections of IBC personnel and their professional obligations, majority opinion (71%) once again disclaims a linkage, but the comments both pro and con merit attention. The following remarks were conveyed by three senior professors: "Scientific inquiry must be open to all ideas. Democrats are more broad-minded, more interested in libertarian values, than Republicans." "Academics are Democrats, and the peer pressure to conform is enormous. So there is a connection." "I'm a Democrat because, like many faculty, I come from a working-class neighborhood. Science isn't involved." College professors might place themselves in the Democratic party column for any one of several reasons, consistent with what we know about relevant sociological and political influences. These respondents demonstrate this point graphically, one emphasizing ideology, another group dynamics, and a third family background. Even our IBC political independent is in lock-step with convention, citing his "belief in the best man" as dictating nonidentification. Here we have the "Mugwump," civic-duty turn of mind which intellectuals, who very often see party in-fighting as "dirty," sometimes exhibit.[36] But, as we have already seen, our cloners, whether young or old, are not nearly so establishmentarian.

A second significant indicator of political commitment is that of ideological orientation. While the average person has no great use for abstractions either in candidate selection or in opinion formation, he is able nonetheless to distinguish between the concepts "liberalism" and "conservatism." Moreover, even among these John Q. Publics, Democrats are disproportionately liberal, Republicans are disproportionately conservative, and independents are disproportionately middle-of-the-road.[37] Still, these terminologies and symbols remain largely the stuff of intellectual discourse, and we would certainly expect our inter-

36. The "Mugwump" mentality in contemporary politics is discussed in James MacGregor Burns, *The Deadlock of Democracy* (Englewood Cliffs: Prentice-Hall, 1964), 211–12.

37. Flanigan, *Political Behavior*, 95–96.

viewees to be more than conversant with them. Indeed, the literature not only finds the typical college professor to be a practicing Democrat; it also confirms his plurality status as a self-described liberal.[38] One would hypothesize that academics, teaching and writing in the liberal arts, are highly sensitive to such themes as social justice, equality of opportunity, civil rights, and economic egalitarianism. Their ideas, I submit, were forged in the crucible of the New Deal; hence, they code themselves as liberals for much the same reason that they code themselves Democrats. And yet, as we saw, gene splicers do not consider themselves Democrats. A check on their political ideology, then, should help us better understand exactly what they mean when they label themselves independents. But one would anticipate that while our young skeptics cannot accept what they perceive as the party of Lyndon Johnson and Jimmy Carter, they have been absorbed sufficiently into the system to call themselves "liberal" rather than "radical." One would further anticipate that our independent senior faculty lean toward "Mugwump" inclinations, relating to "good government" liberal causes but steering clear of partisan machinations.

Recombinant DNA scientists did select liberalism as their most popular ideological tag by far, but the option garnered barely a majority (53%) of total sentiment. Not that conservatism proved an attractive alternative, for its only spokesman was our single Republican. Rather, the large minority split among three postures: middle-of-the-road (21%), "other" (11%), and "don't know" (11%). I can at least say with confidence that Democrats do indeed prefer liberalism by an overwhelming count. And the comments ring true: "I am a liberal Democrat, 1950s style," said one full professor. "Call me an ideological liberal Democrat," noted a junior faculty member, "but I'm not the activist type." So self-identified liberal party members apparently see themselves in opposition to the protest movements of the sixties and seventies. But I was quite wrong to think that independents would flock to the liberal camp; actually, as many as two-thirds disapprove the term in defining their political philosophy. The only reason our liberal contingent achieves majority standing is because resident aliens place themselves in the fold. Approximately 42 percent of cloners do not feel comfortable as a member of any partisan or ideological cohort. What are they, then?

In choosing the "middle-of-the-road" designation, several uncommitteds volunteered definitions of the phrase. These researchers, it turns

38. Ladd and Lipset, *Divided Academy*, 26–27, 158. Among scholars in the "hard sciences," molecular biologists trail only physicists and biochemists in their affinity toward liberalism. Ibid., 64.

out, advocate fiscal conservatism but personal or humanist liberalism. I suspect they are remnants of what used to be called the Eisenhower-Rockefeller wing of the Republican party, but they cannot bring themselves to affiliate with a G.O.P. that has Ronald Reagan as its national leader. Our other nonidentifiers, however, march to their own political drumbeat, such as it is. One philosophical "other" stated: "I am a pragmatist, but a better word might be hedonist." Another self-classified "other" commented: "Traditional categories have no meaning for me. I want to help people in an ethical and responsible way." And a "don't-knower" said: "I move from issue to issue, deciding each the best I can." But this is the same interviewee who had never voted. I am not as yet prepared to generalize about this subset.

By a vote of 58 percent to 42 percent principal investigators disclaim any common linkages between political ideology and vocational interests, but this margin is far closer than the 85 percent to 15 percent negative judgment resulting from the test of comparison between party identification and professionalism. Furthermore, the factionalism reported here seems to spotlight cutting edges of opinion. First, every Democrat-liberal says there is indeed an attitudinal bridge. To aggregate several points of view: "Liberals like to search out new ways of looking at things; conservatives either don't want to rock the boat or they are goal-oriented." Secondly, every middle-of-the-roader said there was no nexus, as did the conservative Republican and 75 percent of liberals who called themselves independents.[39] I see little reason, in theory or logic, why science must or must not express or represent some conception of social change or some perspective on the balance of power. Evidently, though, liberal Democrats see their world and structure their "individual constitutions" this way. Others, who also meet minimal standards of coherence and symmetry in their political profiles, do not share this unidimensional focus.

If these observations are valid, they ought to be prevalent as well among IBC chairmen. Recall that our watchdogs are all Democrats, save one. When asked to select an appropriate ideological label, the terms "liberal," "middle-of-the-road," and "other" received approximately equal support. And when requested to judge whether a link existed between political philosophy and work responsibility, 80 percent fell into the niches established above. For the liberal Democrat, science affords a mechanism for expressing, and bearing witness to, heartfelt social and political needs and values. Such is not the case with respect to researchers who consider themselves either one but not the other, those who straddle

---

39. One liberal-independent answered "don't know."

the ideological fence, or, I would speculate, investigators who have the temerity to consider themselves either Republicans or conservatives.

With ideological parameters on the table, let us return to the variable of academic rank, our surrogate measure for age and, perhaps, for the degree to which cloners have been absorbed into the norm structures of the educational establishment. I thought Democrat senior professors would be liberals, and 75 percent of them were. I also saw no reason to doubt that the Democrat junior faculty would be liberals, which all of them readily admitted they were. But I said the junior independents would also tend to label themselves liberals, thus rejecting a "far left" tag. To be sure, radicalism fails to achieve any support at all, but, putting aside resident aliens, middle-of-the-road politics garners greater enthusiasm than does liberalism itself. There is evidently a group of youthful researchers—as there may well be a larger group of youthful citizens—who cannot identify with either party or either ideological wing, but who consider themselves well within the mainstream of political orthodoxy. And, as I argued earlier, these scientists might well be receptive to a new Republicanism, or, as some might say, to a resurrection of the old Republicanism. The upshot is that I can find only one junior researcher who even vaguely resembles in political profile a "skeptic of the left," to invoke Gerald Pomper's terminology. Lastly, I felt full professor nonpartisans would prove to be "Mugwumps." There is some evidence to support this prediction, as 50 percent of our subgroup chose the badge of liberalism. However, older "Mugwumps"—like younger skeptics (whether apostate Democrats or Republicans)—are informed, articulate spokesmen; and I find it hard to believe that members of either set would confess ignorance when asked if their ideological and professional goals intersected, a configuration which did, in fact, occur.

We are left with a residue of respondents—more than 50 percent of U.S. citizen independents and 30 percent of the entire sample of recombinant DNA researchers—for whom political doctrines and relationships comprise a universe of shadows. A minority of IBC leaders also fits this characterization. Almost to a person, both subgroups eschew membership in a political party; they are unable to articulate a nexus between their notion of partisanship and their notion of science; they either advocate deviant political ideology or advocate none at all; and they fail to discern any relationship between these philosophical commitments, whatever they may be, and professional goals as personal leitmotifs. They also tend to make inordinate errors in assessing recombinant DNA exercises as constitutional phenomena. By "inordinate error" I mean either that replies are in hopeless contradiction to one another or that subjective public policy appraisal is blandly tendered where objective, "constitutional" appraisal is called for.

Some illustrations are in order. When asked what legislative constraints government had the power to place on his research, assuming he were subsidizing his own experiments, one principal investigator said, "I don't think there should be any laws." Another scientist, when queried as to the sorts of strings which the state, as funding agent, could require him to observe, replied, "Any strings it wants, because the money isn't mine"; but this researcher then argued that government, if it wished to support genetic studies, was obliged to support cloning. The attitudes pervading these two subgroups also betray an equivocation, if not a sense of confusion, missing from the general run of respondents' perceptions. For example, with respect to whether scientific inquiry is entitled to First Amendment coverage, one scientist said, "I don't know; it's an issue for lawyers," while another said, "This is an unreal question." I most certainly do not argue that all other gene splicers and committee chairmen present totally consistent sets of convictions or demonstrate exemplary command of constitutional policy items, some of which would be thorny even if placed before acknowledged authorities. But I do say that those in the majority appraise themselves as orthodox "politicals" and, furthermore, that respondents who cannot or do not share that sense of belonging have more difficulty perceiving their own place in the scientific community within some coherent, holistic, articulatable vision of constitutional reality.

In fact, there seems to be two fairly distinct species of deviant political animals contained in this somewhat alienated minority. The first kind tends to be a rather glib, opinionated junior professor with a Ph.D. who teaches in a liberal arts environment and has developed a highly personalized value orientation linking professional code with public philosophical propriety. These scholars are well informed within the intellectual and attitudinal parameters they consider meaningful. They also are better described as cynics than as skeptics. The "enemy," for them, is the System, which is considered either venal, or catering to mediocre standards, or overly impressed with its own place in the scheme of things, or insensitive to the public interest, as the case may be. These practitioners appear to be so determined not to allow perceived systemic evils to tarnish the larger norms of right conduct that they emerge as somewhat aconstitutional when assessed from the theoretical perspective used here. And so personal freedoms are called absolutes, or security blankets, or rationalizations justifying license, or judged inapplicable to the "grubby research rat-race."

The second kind tends to be a not-very-opinionated senior professor who has either earned an M.D. or holds a teaching appointment in medicine. This individual is an old-style independent, in the Michigan survey research sense of the term: uninformed and, with some, uncaring about the societal implications of cloning. There is a literature highly critical of medical education in the United States, arguing that even the better institutions emphasize vocationalism and technocratic concerns to the detriment of both broader intellectual perspectives and deeper cultural forces. I see no need to adopt those charges as my own; many medical doctors and other faculty members in medically oriented disciplines—at least as many as those of whom I am talking—demonstrated in my discussions with them sufficient knowledge, consistency of attitude, and appreciation of these sometimes elusive, larger conceptions. Nonetheless, I would argue that cloners who have gone through Ph.D. programs and who do research within a context of liberal arts study are somewhat better able to structure the inevitable relationships between the world of scientific inquiry and the human elements governing and influencing that world than are academics who lack these advantages.

Again, I should not underplay the major themes. To cloners, liberal values evidently provide the salient political (though not necessarily economic) focus, well surpassing Democratic party allegiance in significance. If there is a single ideological strand which underlies Asilomar, the NIH guideline strategy (though not necessarily the manner in which that grand strategy is implemented), and adherence to the notion of science as constitutional right, then that strand appears to be the liberal commitment. To watchdogs, partisanship is the more intensely held attitude, and it provides a greater sense of political orientation and coherence. We are talking here of team players and organization men, older and, perhaps, more prudent heads. No less than gene splicers, they endorse the public policies that have governed recombinant DNA politics and perceive the new science as free expression. For them, the political system—"their" system—has provided appropriate accommodations, and it is the party that ignites the system. Academics who are first innovators and secondly creatures of political routine can rely on no such norm structures and expectations for meaningful cues; they must fall back, when forced to enter the world of national governmental decision-making, on their gut ideological preconceptions. In the United States during the Jimmy Carter–Ronald Reagan years, it is Democratic partisanship for insider watchdogs and liberal ideology for outsider cloners that best characterize political mood and instinct.

It would be very nice and tidy if I could illumine yet further significant linkages between respondents' political attitudes and their reactions to the various policy questions which compose the recombinant DNA debate. It would have been just as nice and tidy if I had found closer ties between respondents' constitutional values and their reactions to that debate. But I have been discussing in this chapter several disparate themes or subject-matters which come together under one roof only because the entire dialogue has taken so many twists and involves so many aspects of the American cultural and social ethos. Properly understood, the recombinant DNA debate forces us to address these inquiries as though they were pieces of a jigsaw puzzle possessing not one solution but several competing solutions, each standing on sets of fundamental conceptions regarding the nature of this country's Living Constitution. We must not expect too much, then, from our interviewees. In the present context, they are doers, not scholars of either the political circumstances or the political consequences of that doing. However, we have a right to expect from them premises suitably contrived and articulated, reasoned choice-selection, and consensus on all but the murkiest points. By and large, they have not failed us or themselves. Rather, they have provided an indispensible source of fact and attitude respecting the dynamics of their professional universe and the meanings it conveys to them. It remains for us to take these pieces and integrate them into our jigsaw, the better to evaluate the place of science—the place of cloning—in our body politic.

5 Recombining Constitutional Images

On February 11, 1981, biologists Allan Campbell and David Baltimore informed ORDA director William Gartland that they would petition RAC to convert the NIH guidelines from mandatory constraints into recommended procedures. And so, it seemed, the recombinant DNA debate had come full circle. Cloning had spawned no great tragedy for humankind; deregulation was proceeding with ever more confidence and expedition. Perhaps the "purists" regretted Asilomar, or perhaps they had just had enough of bureaucratic turnstiles and incrementalism. In any event, the time seemed ripe to confront, at long last, the basic legal/political question on the gene-splicing agenda: should the NIH as grand giver of research funds continue to *govern* the ways in which authorized experimental designs were to be carried out? We shall see, however, that this issue is but the most dramatic instance among other instances which lead us to ponder the wisdom of state policy toward DNA manipulations. The central question, then, is one of normative constitutional theory, involving matters of institutional arrangement, of civil liberty, and especially of attitudes and expectations regarding the role of science, the art of power enhancement, the definition of public interest. For the Constitution means many things to many people; its themes and moods sometimes overlap or even appear contradictory as historical trends unfold, as exigencies arise and trigger unprecedented responses from the citizenry and those it has chosen to lead. How to weave these disparate images into a new constitutional profile so as to meet the many challenges of genetic engineering, while at the same time keeping faith with yet deeper patterns of right political conduct, is the central theme of this chapter.

The NIH Role: To Orchestrate or Not to Orchestrate

The Campbell-Baltimore proposal, as presented to RAC for discussion in April of 1981, would have accomplished the following: (1) All NIH compulsory requirements contained in the recombinant DNA guidelines and the organizational structures sustaining them would cease to have binding force or effect. (2) A voluntary physical containment standard of P1 would receive an NIH imprimatur as the generally accepted safety criterion. If unusually volatile materials were employed, then investigators would be urged to utilize precautions normally considered appropriate for those materials, whatever the research context. (3) The prohibitions would stand as before, and RAC would continue to provide exceptions to these, exemptions from the new guidelines, and consultative channels for large-scale commercial producers.[1]

The level of consensus at this meeting was probably higher than outsiders might have anticipated.[2] Nobody argued that cloning had thus far exhibited real hazard, nor did anyone believe the restrictions, as then constituted, deserved to retain their present form, much less should they be toughened. In fact, the realization was expressed that deregulation *ad infinitum* absent a review of mandatoriness would render the administrative artifice an empty shell and have the effect of making NIH, and its board of scientific advisors, look rather foolish. But, as before, there was division over *political* issues. If we can't undo emergency actions once emergencies abate, then perhaps scientists will be less flexible the next time around, Norman Zinder implied. To this, Allan Campbell added that RAC's "function is to deal with danger [not] fear." But public trust was on the minds of many. Precipitous action, uninformed by strong support from concerned interests, could create a powerful backlash. At the least much more was needed in the way of exposure and input than would follow from routine notifications through the *Federal Register*. And there was the usual discussion about public and participant attitudes respecting the recombinant DNA controversy, followed by not only the usual assertion of unconfirmed conclusions but also the usual exasperation because nobody really knew what significant groups thought. For example, some advocates of voluntarism argued that the average person holds no fear of cloning research. As a matter of fact, opinion polls show that the man in the street strongly opposes such investigations.[3] From the other side came the view that if a strong minority of

1. *Fed. Reg.* 46 (March 20, 1981): 17995.
2. See National Institutes of Health, Recombinant DNA Advisory Committee, Minutes of Meeting, April 23-24, 1981, 8-15. Hereinafter referred to as "Minutes of RAC Meeting."
3. Jon D. Miller, Kenneth Prewitt, and Robert Pearson, *The Attitudes of the U.S.*

practitioners frowned on the Campbell-Baltimore proposal, then "the scientific community invites societal control from without." However, my data indicate that a majority of cloners oppose any such drastic upheaval as that being initiated, but whether this sentiment could generate any spinoffs in the form of outside interventions would depend on political conditions then unknown and perhaps unknowable. When the time came to send the recommendation to subcommittee, the skeptics had mustered enough support to fashion for the minipanel the broadest of instructions: consider whatever issues you think important and make whatever proposals for reform you think warranted.

The working group, appointed by RAC chairman Ray Thornton, consisted of thirteen members. Its makeup was nicely balanced indeed: those advocating sweeping change could expect Norman Zinder to represent their views; those highly suspicious of change could count on Harvard's Richard Goldstein to make their voices heard; those searching for intermediate solutions had a vigorous spokesman in Susan Gottesman. As matters unfolded, the group's most important contribution was a paper entitled "Evaluation of the Risks Associated with Recombinant DNA Research."[4] In this document, the subcommittee concluded, after a careful examination of the evidence and without a dissenting opinion, that only experiments conducted with the specific purpose of expressing alien genetic material designed to work change in the host organism's functions posed any meaningful chance of risk. Hence, P1 constraints—which were far more prudent than the safety standards historically considered necessary and proper—certainly would suffice for all other recombinant DNA maneuvers. With these scientific givens, major guideline surgery seemed an inevitable and highly desirable consequence.

As to the political tools to be utilized in effecting these changes, however, much disagreement surfaced and proved incapable of resolution. A majority of eight was willing to go even beyond Campbell-Baltimore on the containment question, not only setting P1 as the appropriate benchmark for all combinations of nonpathogenic materials whether cloned in prokaryote or eukaryote recipients but also eliminating the prohibitions as a discrete class. Any experiment involving substances or accomplishing objectives heretofore proscribed (e.g., utilizing potent toxin genes, mounting many large-scale investigations) could proceed in the manner of other recombinant DNA endeavors involving pathogens. And those standards, per Campbell-Baltimore, would be the safety levels assigned

Public toward Science and Technology (Chicago: National Opinion Research Center, University of Chicago, 1980) as cited in John Walsh, "Public Attitude toward Science Is Yes, But—," *Science* 215 (January 15, 1982): 270.

4. *Fed. Reg.* 46 (December 4, 1981): 59385–90, 59392–94.

by such agencies as the Center for Disease Control to experimentation employing the subject materials outside the gene-splicing context. But the report retained the principle of mandatory enforcement, while at the same time deleting the requirement for community representation on the IBCs. It was "foolhardy," the majority wrote, to dismantle a ready mechanism for independent review and oversight, especially when the result might entice other political groups to enter the vacuum. No explanation was tendered for revising the makeup of the campus panels, bearing witness probably to the fact that the committee wanted to avoid political issues while, at the same time, acknowledge *sub silentio* political considerations.[5]

A five-person minority, representing the Campbell-Baltimore orientation, agreed that the prohibitions could lapse in the light of shared scientific interpretations. And they noted with pleasure the finding that recombinant DNA research contained no dangers to persons or environment beyond those normally associated with the particular genetic substances and host organisms utilized. From this, they saw little reason not to defang the NIH controls as coercive tools, for the contrary recommendation could only result in administrative "capriciousness." Two of these dissenters, Norman Zinder and Edward Adelberg, would have gone further and simply abolished the guidelines altogether. At this point, the restrictions could be justified only on "social and political" grounds, criteria of no concern where scientists could see clearly the absence of physical danger.[6]

When RAC convened in September of 1981 to discuss the working group's document, no one expected a quick and easy resolution of these differences. Rather, the plan was to vote on amendments, disseminate the final package to as broad a group of interested parties as seemed feasible, process the feedback, and meet once again for final disposition of the matter. A review of the ensuing discussion[7] shows that none of the protagonists introduced new and dramatic evidence to augment their litanies of previously cited convictions. As the debate was essentially political in focus, one would have expected the working group to gain the upper hand; its recommendation seemed well tailored to occupy a middle ground between the status quo and voluntarism. Somehow, though, David Baltimore seems to have done a good job of seizing upon the indecisiveness among his colleagues, and *his* proposal, as he presented it, emerged as the appropriate middle position, balancing on the one hand the Zinder-Adelberg notion that scientific judgment should dictate

5. Ibid., 59383, 59390–92.
6. "Minutes of RAC Meeting," September 10–11, 1981, attachment IV, 2.
7. "Minutes of RAC Meeting," September 10–11, 1981, 3–15.

guideline abolition and on the other hand the subcommittee recommendation that cumbersome, obligatory review by IBCs was still a practical, though scientifically indefensible, necessity. Politics *were* important, Baltimore admitted, but centralized advice and expertise carried on through the good offices of RAC and ORDA were certainly sufficient to satisfy forces still sensitive to public accountability. If the campuses wanted more, they could mount their own preclearance and monitoring programs. He also supported recommendations, to be listed among his proposed constraints, warning against the deliberate tampering with microorganisms so as to attenuate their resistance to drugs in a manner that might compromise the utility of those substances as disease-fighters and the deliberate use of DNA from highly lethal poisons, including the botulinum and diphtheria toxins. When the dust settled, RAC had gone on record, 16 to 3, as favorable to a revised Campbell-Baltimore package, consisting essentially of P1, P1-LS containment, the two prohibitions noted above, disease-control criteria for all other pathogenic agents, and advisory status for these several caveats.

RAC's resounding endorsement of the guidelines as a code of good laboratory practice and nothing more was published in the *Federal Register* on December 4, 1981, precisely at the time my interview schedule was drawing to a close. But, of course, the news of RAC's bold thrust had spread quickly, and both the cloners and the watchdogs to whom I talked well understood that in answering several of my questions they were participating in a kind of referendum on the recombinant DNA debate as it was then structured. Especially relevant is that both groups favored retention of the principle of obligatory review, and that as many as three-quarters of the researcher sample felt this way. What impact, if any, would these opinions have on public policymaking?

Actually, the counterrevolution began on the very day of David Baltimore's Bethesda triumph. RAC chairman Ray Thornton, a lawyer and the president of Arkansas State University as well as a former congressman and key participant in the 1977 House debates on proposed legislation to restrict gene splicing, having presided over these difficult deliberations with the necessary objectivity and neutrality, now filed a statement taking vigorous exception to the committee's position.[8] He viewed with alarm the notion that political and social questions could be ignored when scientists take research money from the people's government. Investigators can no more isolate themselves from society than society can isolate itself from the experimentation it fosters, he submitted. If work with human subjects ought to be regulated, then so should the deliberate release into the environment of potentially

8. Ibid., attachment V.

dangerous organisms, a practice, incidentally, which RAC had not even been willing to condemn as professionally unacceptable in its meeting that day.[9] Public participation, as exemplified by RAC processes, was important in the present context, he believed, for two reasons: (1) human experience demonstrates that research capable of great good is also capable of great harm; (2) the public has a vested interest, given its financial stake as subsidizer, in any "dangerous or *ethically or socially offensive experiments*" (italics mine), and the public will find a way to regulate these practices no matter what NIH ultimately decides to do on the voluntariness issue.

That these attitudes could be ignited into action was made manifest when, on December 7, 1981, there appeared in the *Federal Register* an alternative proposal for deregulation.[10] Its author was Susan Gottesman, an NIH molecular biologist and chairman of the recently disbanded RAC working group. Essentially, her recommendation retained the entire NIH and IBC coercive machinery and lowered, though not to the extent advocated in the revised Campbell-Baltimore RAC-approved option, sundry containment provisions. More specifically: deliberate release into the environment of recombinants, deliberate alterations of microorganisms with respect to drug resistance (as noted in the Baltimore proposal), and deliberate use of various toxins lethal to vertebrates would no longer be prohibited, but they would need RAC review, NIH certification, and IBC preclearance; deployment of CDC Class 4 or 5 organisms as well as all work in excess of ten liters would also no longer be prohibited, but these would require IBC prescreening; experiments involving other pathogens and the insertion of alien DNA sections into animals and plants would now also be a matter of IBC censorship rather than NIH censorship; studies utilizing nonpathogenic prokaryotes and lower eukaryotes, most of which required campus prior restraint, could now commence with mere notification to IBC personnel and be performed under P1 conditions, though a postreview process would not thereby be waived.

Strongly reflecting RAC sentiment, ORDA made a concerted effort to disseminate these competing blueprints; copies were mailed to more than 4,300 interested persons and groups. Reactions to both proposals as well as new approaches and compromises were encouraged. A total of ninety-five letters ultimately reached Bethesda headquarters on this subject and, as collated some time later by NIH staff, fell into the following categories: (a) three communications endorsed the Adelberg-Zinder formula calling for immediate guideline dissolution; (b) thirty-five letters sup-

9. "Minutes of RAC Meeting," September 10–11, 1981, 13.
10. *Fed. Reg.* 47 (December 7, 1981): 59734–37.

ported the RAC-Baltimore option; (c) ten commentators tried to strike a balance between the Baltimore and Gottesman positions, some wanting voluntary enforcement coupled with stronger constraints, and others opting for mandatory enforcement coupled with looser criteria; (d) thirty-two reviewers advocated the Gottesman proposal; (e) ten respondents, including the chairmen of the Harvard University IBC, the Harvard Medical School IBC, and the Boston Biohazards Committee, opposed both formulas for revision, and requested NIH to stand fast on its regulatory strategy; (f) one reply, representing the views of three congressmen on the House Committee for Science and Technology, expressed particular displeasure at any thought of dismantling the IBCs, which they saw as a significant forum for "public review."[11] It is fair to say that when RAC finally sat down to hammer out its official recommendation to NIH, approximately half of all the responses in hand favored the Baltimore package, while the other half supported either the Gottesman package, the guidelines as they stood, or something in between the two.[12] Clearly, the impending confrontation could not be resolved satisfactorily by any straightforward reading of these raw tallies. On the other hand, the lack of consensus portended a much closer battle than the 16 to 3 vote count of September.

Perhaps the most important meeting RAC has held to this date took place in February of 1982.[13] At once, Chairman Thornton seized the initiative, announcing an intention to state his personal view before he retired from the fray as presiding officer. The guidelines, he argued, were neither laws, nor regulations, nor an informal code. All of this was for the good, because statutes are difficult to formulate, administrative rules can be changed only through formalistic procedures, and voluntary standards provide neither a conduit for communication between policymaker and scientist nor a vehicle for public participation. Donald Frederickson, one of the architects of the NIH restrictions and until a few months previously its director, had fought the good fight to keep cloning research both on the move and on the straight and narrow, Thornton implied; this was not the time to forget the wisdom of his contributions.

David Baltimore then took the floor and immediately assumed a defensive posture. Some critics wanted to retain the IBCs, he began; certainly a specific statement on that score could be appended. Others feared deliberate release of mutants, he said; of course the revisions

11. *Fed. Reg.* 47 (April 21, 1982): 17173–76. Nine of these letters arrived after the due date and could not be considered on decision day. Four other communications dealt with special considerations irrelevant here.

12. "Minutes of RAC Meeting," February 8-9, 1982, 8.

13. Ibid., 5–19.

might easily plug that loophole with explicit condemnatory language. A few observers had even accused him of having ties with a commercial outfit, Collaborative Research of Waltham, Massachusetts, he noted, thus compromising his position as a neutral decisionmaker. In fact, Baltimore rebutted, he had made no secret of these ties, and if anything, that role would counsel a prejudiced party to favor retention of the guidelines because there was some danger that the Greater Boston community might impose yet stronger regulations than heretofore should NIH depart the scene.

Others now joined in a comparison of opinions. The more they talked, the more it became clear that the dominant preoccupation was "public consent." Said one member: if the correspondence we received shows anything, it is that the public is not prepared for recombinant DNA guidelines as mere recommendations. Said another: legislators don't really care about scientific assessments, only public reaction; hence, RAC must tread warily in this unstructured domain. And when David Baltimore reported that, in his firm belief, a majority of cloners saw their labors as no more dangerous than other "mainstream" biomedical endeavors, Patricia King, a Georgetown law professor, shot back that the public could hardly verify his hunch by cross-examining those who had taken the trouble to inform RAC of their heartfelt opinions. In fact, my data suggest that Professor Baltimore's characterization of gene-splicers' sentiments was probably accurate, but by no means does it follow that these researchers advocated his blueprint for deregulation. The only point Baltimore's antagonists would concede was that the Gottesman option was too restrictive, vesting in the federal bureaucracy more authority than was appropriate to mete out P2 and P3 requirements. The result was a motion to accept her plan, coupled with an amendment which insisted upon an expeditious review and simplification of these rules. By a crushing 17 to 3 margin, the panel adopted that proposal, hence repudiating its previous commitment to voluntarism.

But RAC could not quite let go of the matter, could not quite reconcile itself to the implications of what it had done and why. Having acted politically for purely political motives, some members recoiled from the thought that others might also act politically for purely political motives. Let us go on record as opposing more stringent state and local laws, a panelist suggested. No, another responded, that would be inflammatory, because, as a member of the Boston Biohazards Committee, he could see a growing conviction to include within these restrictions other provocative modes of experimentation. Well then, came the argument, why not encourage these boards to branch out; at least cloning would lose its undeserved solitary second-class status. Good heavens, was the rejoinder, that would take us back to 1976! Finally, a motion was introduced declaring RAC's belief that the NIH guidelines were based on the best scientific evidence, and that communities should not escalate these criteria unless they present "credible" data demonstrating "unique risk." But suppose we incite local politicians to contemplate more elaborate regulations, a member inquired, because they perceive our call as a form of elitism, even arrogance? By a vote of 9 to 6, the motion was tabled. The debate had ended. Approximately ten weeks later, a notice appeared in the *Federal Register*; Donald Frederickson's interim successor for RAC policy oversight, Bernard Talbot, had approved the Gottesman package.[14] The mandatory "guidelines" remain in place.

Nonhuman Genetic Engineering: How to Regulate

From Asilomar to 1985, the preeminent building block of recombinant DNA discourse and decision has been politics. Politics, like scientific or even artistic activity, is essentially neutral in cast and character; like cloning itself, politics can be a tool for achieving any number of purposes, results which very often, of course, are not so neutral in their social implications. But under the Constitution, politics is supposed to transcend neutrality in the sense that the basic rules of public policymaking have achieved standing as "higher ground" or "according to Hoyle" or "*due* process." To the extent that these norms *are* coded as neutral principles, that perception merely enhances their "goodness," for neutrality so defined bespeaks evenhandedness, objectivity, and freedom from the taint of "result orientation."[15] It is this notion of constitutional value, of process considered due, which holds the key to assessing the ways we have treated cloning as a burgeoning political phenomenon. Putting aside issues pertaining to the especially sensitive realm of human genetic research—these deserve, and will receive, separate treatment—we turn first to the normatives surrounding government supervision of all other DNA-splicing exercises.

The Constitution intersects these cloning endeavors on several due process planes. For one thing, there is the notion of recombinant DNA research as free expression, the theory that our constitutional law, as a body of knowledge and juridical expectations, provides shelter to the "speech" aspects of gene splicing; that these experiments, in short, are in the nature of substantive liberty and, hence, the researcher's right to perform them can be circumscribed only through appropriate standards of

14. *Fed. Reg.* 47 (April 21, 1982): 17172-76.
15. Herbert Wechsler, "Toward Neutral Principles of Constitutional Law," *Harvard Law Review* 73 (November 1959): 1-35. Cf. Carl B. Swisher, "The Supreme Court and 'the Moment of Truth,'" *American Political Science Review* 54 (December 1960): 885.

fairness. We are talking here only of theory, to be sure, but constitutional theory is the mainspring of constitutional decisionmaking; one day, the courts will have to develop some link between the Bill of Rights and contriving these new life forms.

Next, there is the strong support accorded this nexus, this conception of the rules, by both scientists in the field and public representatives in the field. Both cloners and watchdogs advocate the proposition that recombinant DNA manipulations trigger free expression considerations, because these efforts, at least as academic ("pure") explorations, constitute searches for biological knowledge. If theory is to have its day outside the pages of law books and the arguments of pedagogues, then those who practice the art and those who represent society's first line of defense against abuses of that art ought to share a basic regard for the core values of that theory. Indeed, Americans as a whole oppose restrictions on scientific inquiry unless special circumstances obtain, and 50 percent of "attentives"—i.e., the better educated, more interested stratum—favor further development of the new modus operandi.[16] We need not fret because most people favor a ban on cloning, because most people also favor a ban on speech by fascists and communists. American constitutionalism does not require the endorsement of *vox populi* through Gallup-style polls.[17] But to the extent that recombinant DNA experimentation as substantive freedom does require support from the general public in order to become an operational value, that support seems clearly to exist.

Third, there is the general acceptance, expressed by gene splicer as well as chief campus biosafety official, that the NIH guidelines have comported with constitutional criteria. That is, the standards are perceived to be reasonable, given prevailing scientific and political considerations. The state, for them, has shown a capacity to meet unique research instruments and goals with measured political instruments of control.

There is one other relevant due process intersection, and it involves the politics of regulation within the larger recombinant DNA debate. Cloners support the spirit of Asilomar, even in retrospect, and they support the principle of mandatory constraints in the NIH guidelines. They also favor national safety standards. The problem is that they are at sixes and sevens over Bethesda's oversight model. What are the constitutional parameters of this problem?

16. Walsh, "Public Attitude toward Science," 270–71.

17. For the latest research update on public opinion surveys respecting the civil liberties attitudes of Americans, see David G. Barnum, "Decision Making in a Constitutional Democracy: Policy Formation in the Skokie Free Speech Controversy," *Journal of Politics* 44 (May 1982): 480–508.

I have noted the importance of public and interest group attitudes as living constitutional material; the issue before us involves norm implementation as living constitutional material. In other times, the state passed laws and people were expected to live up to them or face punishment. The theory of the modern-day administrative state is that highly technical problems require highly technical solutions. The Progressives, who championed the idea of expertise as a substitute for politics, never envisioned a "constitutional" order in which highly technical solutions carried with them tenuous enforcement procedures. And yet, NIH establishmentarians have contended, law becomes palatable when implemented in-house by guideline rather than out-of-house by regulation. One of the problems with genetic engineering guidelines, though, is that the practitioner knows infinitely more than the bureaucrat. After all, we are not talking here of income tax enforcement, where the IRS far outstrips in knowledge the wage earner, or drug safety, where the FDA can muster as much expertise as the pharmaceutical firms. Therefore, these scientists are not going to be sold on paperwork rituals and responsible monitoring procedures unless they can be shown either scientific or political need. As I have indicated, the question of scientific danger may have prompted Paul Berg to stay his hand early on, but it has not been the cornerstone of gene-splicing controversy. As for politics, the research community would like to do only research and leave the questions of public acceptance and public consequences to others.[18] In fact, one can argue that if scientists are going to parley over politics, they may well lose the status card they need to flourish as an "apolitical elite."[19] But, under current conditions, they do need money from NIH. So the message they get is this: "Just keep your noses clean and play along; in return, we'll do our best to keep the grants coming, we'll deregulate as soon as the political climate lets us, and, in the meantime, we promise a minimum of oversight." Years ago, Bradford Gray, in his study of Bethesda's human subjects guidelines, pointed to the seeming lack of an effective surveillance program, without which social control and accountability were being defeated.[20] NIH has not learned very much in the interim. And so law becomes quasi law, and as much as half the recombinant DNA community, expecting law really to be law, throws up its hands in disillusion-

18. Joseph Haberer, *Politics and the Community of Science* (New York: Van Nostrand Reinhold, 1969), 1.

19. Robert C. Wood, "Scientists and Politics: The Rise of an Apolitical Elite," in Robert Gilpin and Christopher Wright, eds., *Scientists and Rational Policy Making* (New York: Columbia University Press, 1964), 44.

20. Bradford H. Gray, *Human Subjects in Medical Experimentation* (New York: Wiley, 1975), 14.

ment and cynicism. That is what some cloners are driving at when they blame flawed regulatory policy on "the system."

Of course, if scientists knew a little more political theory, then maybe they wouldn't fall victim to the "redburn" syndrome. Certainly they have established their reputations as thinking liberals, who favor broad government initiatives to promote egalitarian social policy. Can we really expect them to know also that Lyndon Johnson's liberalism was not Franklin Roosevelt's liberalism,[21] that, as I pointed out in Chapter 3, delegation of power run riot and interest group self-government have imparted to "law" an unprecedented vagueness even with presidents like Richard Nixon at the helm?[22] If researchers don't relate to political bargaining—except perhaps as a matter of survival when playing the grantsmanship game—then it is hard to see them relating to treatises on bargaining theory, whether these help explain the dynamics of national policymaking or not. What many of these scientists do know, what they can sense, is that NIH guideline enforcement lacks *legitimacy.* What keeps interest-group cooptation of the administrative state, including the Asilomar majority's cooptation of NIH guideline-enforcement strategy, from becoming part of the Living Constitution is its lack of legitimacy as legal, as due, process.

What level of political sophistication, then, can we expect from scientists? More specifically, to what extent must they, must we, rethink the constitutional politics of implementing constraints on cloning research? As a matter of self-interest, genetic engineers need to envision their craft from the standpoint of guaranteed substantive liberty. My data show that they make the necessary connections in the abstract; but it is equally important to bring the principle to life in the form of constitutional practice. In short, the concept of inquiry as experimentation must be *judicialized,* just as other fundamental private rights in this country are becoming ever more judicialized, and for the same reasons. That is, cloning, draped in the attire of protected individual prerogative, aggrandizes to itself the political status necessary to compete for resources, especially the resources of disciplinary integrity and even subsistence, in our pluralistic, group-oriented society. The judicialization process, moreover, has a therapeutic, cleansing effect. The people's represen-

21. See generally Theodore J. Lowi, *The End of Liberalism* (Chicago: Norton, 1969). Even FDR's Blue Eagle program, with its amorphous-beyond-comprehension "guidelines," was enforceable under the federal criminal code. The question of who enforces the NIH penalty provisions and how has been a nagging dilemma. For an update, see Marjorie Sun, "NIH Developing Policy on Misconduct," *Science* 216 (May 14, 1982): 711–12.

22. Typical is the term "national security," Johnsonian-Nixonian jargon permitting law enforcement agencies to do pretty much what they thought the traffic would bear.

tatives know they must regulate in good faith, must establish norm structures consistent with custom and usage, or the courts may well step in and brand their machinations unconstitutional. It borders on theatrical absurdity that the great apostle of "libertarian legitimacy" in recombinant DNA judicial politics should not be the genetic researcher but should be Jeremy Rifkin.

But could these traditional legal modes have processed the politics of recombinant DNA? What form would the scenario have taken? Instead of Asilomar, NIH, ORDA, and RAC serving as instrumentalities of state police power, we could have had a great debate in the halls of Congress. Theories of governmental power would have certainly clashed with theories of scientific right. Witnesses from all corners would have been called, and lawyers would have argued and negotiated in the back rooms. Legislation, spelling out correlative liberties and duties, would have emerged. It might have contained the following: (1) P3 means work conducted under such-and-such obligatory conditions; (2) no contrived organisms shall be released into the larger environment; (3) research employing such-and-such toxins is forbidden. If legal codes cannot define precisely the terms "P3 procedures," "larger environment," and "toxins," then they cannot define anything. All impermissible actions would be punished, as all other impermissible actions at the workplace or on the streets are punished. Restrictions geared to the content (rather than the manner) of experiments, the use of vague regulatory criteria, overbroad constraints lumping together pure and applied investigations where the First Amendment seems to demand demarcation, and reliance upon unduly ambitious censorship mechanisms would lead to old-fashioned courtroom confrontation. A preemption clause could well have gained support, ousting cities and states from the field. To those who might argue, "But Congress gave it the old college try in 1977," I say, "By not recognizing scientific inquiry as implicating constitutional freedom and recombinant DNA research as implicating free expression, Congress missed the central focus of analysis and almost deserved to fail." And to those who might argue, "But science would be slowed down," I say, "If speech can be slowed down, then so can science." Indeed, should practitioners of experimental research, which is "only" quasi speech, be too myopic to rely on judges as enforcers of "higher law" prescriptions, the risk is rather considerable that the system will treat science, including genetic recombination, as mere conduct, a clear and present danger which leaves the apoliticals at the mercy of politicals whenever push comes to shove. It is through these traditional constitutional mechanisms that the process of recombinant DNA regulation becomes truly due.

I pose two final questions under this heading: Who exactly does the enforcing, and how should it be accomplished? There is still a place in our scheme for institutional biosafety personnel who can perform informational clearinghouse functions and even informal monitoring tasks, but in no case do I think campus committees should have ever been delegated censorship or federal law enforcement responsibilities. What we will eventually need, and should have had at the beginning of the gene-splicing debate if not earlier, is a team of research safety inspectors operating out of the Center for Disease Control, the National Institute of Occupational Safety and Health, or some other regulatory agency. These inspectors would make the rounds of all experimental installations, public and private, where hazardous work is carried out, whatever the scientific/technological mission or methodology may be. The term "hazard" must be spelled out in the law; while fear itself, in the form, say, of science fiction scenarios, may be sufficient to justify federal oversight of applied research, it is not enough to constitutionalize federal regulation of quasi speech. And, to reiterate, "times, places, and manner" rules may have to be more tightly drawn for "ways of knowing."[23] I have no data to support, nor do campus cloners support, the allegation that, at this time, industrial, biotechnological endeavors require stricter rules than *hazardous* academic recombinational endeavors. But government inspectors must be educated as to the pure-applied science distinction, not only because in future years commercial enterprises will become riskier but also because these watchdogs must make sure that particular regulatory standards are implemented only against those research exercises for which they were intended. The root of the problem, however, lies not with the state as potential regulator but with the facility as ongoing investigator. Police officers, after all, are expected to master the nuances of search and seizure doctrine; surely, it would be far less demanding for monitors of research to distinguish pursuits of knowledge from pursuits of other things, should the occasion arise. I see no reason, in the abstract, why pure science practiced at the Bell Laboratory's Physics Division should not receive the same free expression coverage as pure science conducted in the Princeton physics department. But if MIT wants the government to meet a higher presumption in regulating and overseeing experimentation as expressive activity, it is going to have to find devices for categorizing rationally that which is, indeed, basic

23. As I have often said, broader restrictions than these can be applied to funded research in the form of conditions, and NIH would not be precluded under my scheme from so insisting; but I do not agree that what the criminal law enforces against other quasi-speech practitioners should, as a matter of policy, be enforced against scientists by the allegedly more palatable but, in fact, the more sweeping string attachment device.

"hazardous" science. It will not do for university or industrial or think-tank program directors to argue that such mapping would "make it impossible . . . to conduct research along an integrated R&D spectrum" and "force [us] to fit [our] research programs into predictable boxes [thus] distorting the work that would be contained in the boxes."[24] To respecify: It is one thing to discover the properties of sundry conductors, and quite another thing to construct transistors for profit. If patent judges can distinguish between utilitarian and nonutilitarian principles, if cloners and campus IBC personnel can agree on what is pure research and what is not, then MIT and Bell Labs and Genentech and MGH can also make allowances for these significant differences. Even government-owned installations are not immune from constitutional checks here, for if federal officials are going to contract out basic science tasks to academics working at Brookhaven, then I refuse to believe, absent language in the agreement itself, that these researchers have bargained away by implication their First Amendment "free science-free cloning" privileges. In such fashion do we integrate the regulation of recombinant DNA experimentation with experimentation generally, bringing together very old and very new theories of constitutional value and expectation to achieve a fair and proper balance of interests.

Nonhuman Genetic Engineering: How to Encourage
In March of 1982, a distinguished group of university presidents, faculty, and industrial leaders met at Pajaro Dunes, California, to discuss and perhaps even effect consensus on a variety of controversial issues regarding cloning for dollars. The conference's format itself resembled features of political DNA recombined from two previous meetings of note, the 1977 NAS forum in Washington, D.C., and the Philadelphia Convention of 1787. As with the former, students, labor, and representatives of environmental and other public interest aggregations were, from the outset, purposely excluded. As with the latter, the doors were closed to the press in order that candid give-and-take might prevail.[25]

Unfortunately, the assembly failed to produce agreements that might challenge in importance the Asilomar accords, much less the Philadelphia compromises. On general themes, the members did concur in the not very startling presumption that campus-corporate bargains were just fine, provided private sector funds could be kept from undermining salient academic values. But on specifics, the results could only be called disappointing and confusing. Should contracts between firm and school

24. David Dickson, "A Battle over Bell Labs," *Science* 216 (June 25, 1982): 1393.
25. These remarks are based on Barbara J. Culliton, "Pajaro Dunes: The Search for Consensus," *Science* 216 (April 9, 1982): 155–58.

be made public? Maybe yes, maybe no. Should universities give corporate donors exclusive licenses for patent development? Maybe yes, maybe no. Should researchers be permitted to develop their own arrangements with business interests? Maybe yes, maybe no.

The problem confronting the Pajaro Dunes participants was the same one that has so often plagued the RAC membership: lack of proper focus. The NIH guidelines have been masquerading for the past several years as largely exercises in scientific judgment; in fact, they are essentially exercises in political judgment. The cloning for dollars facet of the recombinant DNA debate has been masquerading as largely a question of economics; in fact, it is also essentially an exercise in politics. Universities are political institutions, prime movers in the struggle for power among networks of competing influence. The real question, then, is not "How does one behave with money?"[26] It is, rather, "What political function should Harvard, Princeton, and Michigan be performing?" Better yet, because these facilities as idea-centers are little more than their faculties and students, the issue is, "What political function do we want our academic research scholars to perform?" And if politics is the primary concern, then constitutional values, correctly understood, provide the overriding focus.

Americans tend to view constitutional freedoms as a set of thou-shalt-nots. The state is forbidden to deprive us of trial by jury, equal protection under law, and so on, the document reads. Nowhere do the Framers assign to the state an affirmative duty to enhance any of our rights. These duties, if they exist, must be found in tradition, in usage, in patterns of practice. Whether they do exist is a question of considerable controversy. There is no controversy, however, about whether the Constitution authorizes state *encouragements* to the free exercise of rights. Thus, Congress is empowered to enact appropriate legislation eradicating the vestiges of slavery and eliminating state impediments to our enjoyment of due process, whether procedural or substantive. And we can also find encouragements in custom and in precedent. What are references to the Almighty on our coins and in our pledges of national allegiance if not symbolic encouragements to religious exercise? The question therefore arises: What encouragements as a matter of tradition, as a matter of living constitutional practice, has the state accorded scientific freedom?

The most obvious stimulant is the patent grant, if only because it resides in Congress' delegated, discretionary arsenal of policy tools. It is conceivable that William Shockley was thinking "transistor" and thinking "patent" when he published his seminal theoretical pieces, but it

---

26. See the discussion in Marjorie Sun, "NIH Ponders Pitfalls of Industrial Support," *Science* 213 (July 3, 1981): 113–14.

would be hard to convince anyone that monopolistic control over the marketing of artifacts has been a prime spur to innovative research in the natural sciences. There is even considerable question as to its importance, not to mention its desirability, in the world of commerce. As Price has commented, patents are now mostly "defensive tactics"; the key to competitive advantage is engineering technological change through brainpower.[27] In its first two years, the classic Boyer-Cohen patent on gene-splicing processes yielded $1.2 million in fees and royalties, these to be divided equally between Stanford and the University of California.[28] At that rate, each university could anticipate a return of $5.1 million over the seventeen-year life of the franchise. That is a tidy sum, but it pales next to the Whitehead-MIT agreement. Besides, how many recombinant DNA exploits will result in such groundbreaking and, hence, lucrative payoff? Indeed, cloners themselves, as we noted in Chapter 4, have no illusions that these proceeds can make a large dent in defraying their experimental costs. On the question of "general welfare," moreover, we have seen that both Jefferson and Franklin refused to patent their vendibles with that consideration in mind, while Harvard insisted that medically relevant inventions and discoveries be dedicated to the public as late as 1975. I do not argue that the patent system is inherently flawed;[29] indeed, it may well spur the dissemination of ideas. The point is that there have always been doubts about its propriety and efficacy, and that academic-industrial relations need hardly come a cropper of the patent-or-perish horrible if these parties exercise a little care.

With respect to this standard of care, well-informed cloners believe patents developed through faculty ingenuity should devolve upon university officials, but they also say monies accruing therefrom must not be allowed to establish favored and disfavored professional classes. It is to be hoped that few would disagree with that proviso, which seeks to maintain minimal standards of content neutrality in the face of temptations to commercialize. On the other hand, patents currently pose the very distinct threat of putting educational institutions in debt to large benefactors. In the "good old days," academics doing basic science negotiated with industry largely through the medium of individual consulting ties.[30] In our time, Monsanto and Du Pont donate funds hither,

27. Don K. Price, *The Scientific Estate* (Cambridge: Harvard University Press, 1965), 33-34.

28. Marjorie Sun, "Stanford's Gene Patents Hit Snags," *Science* 218 (November 26, 1982): 868.

29. For the argument that Congress should abolish the patent system, see Wassily W. Leontief, "Introduction," in Leonard S. Silk, *The Research Revolution* (New York: McGraw-Hill, 1960), 6.

30. Dorothy Nelkin, "Intellectual Property: The Control of Scientific Information," *Science* 216 (May 14, 1982): 706.

thither, and yon—committing particular faculty to particular research inquiries subject to no peer review—in return for which the firm will receive exclusive licenses from institutional patent holders should these issue. And so enters the spectre of campus trade secrets. To all this, some have said, "It seems unlikely that universities will find better terms [from industry],"[31] and they also say higher education has nowhere else to go. If academic freedom as a vehicle for the idea marketplace means anything, it is that the burden of proof for both these propositions lies with their proponents. Let us examine the evidence from the standpoint of other traditional encouragements.

The most important public sector research stimulant, certainly for gene splicers, is the federal grant. And yet just as the politics of the patent reward system has served to make its living constitutional dynamics shaky, so the politics of the grant reward system has impaired its standing as legitimate norm structure. The United States government has never shown that much interest in funding pure research for its own sake. Vannevar Bush, of course, recognized the problem, but the National Science Foundation has failed even to achieve the high estate he considered a realistic goal. Nor will it suffice to blame "Reaganomics" for today's shortcomings. What the current administration wants to do is get Washington out of programs which make it industry's competitor, not the Guggenheim Foundation's competitor. While the NSF and NIH budgets have fallen victim to belt-tightening, the cause is animus toward domestic spending generally plus increased tax relief. Absent these resolves, argues the White House, budget deficits will be even greater, inflation much higher, and economic growth even slower. As Mr. Bush pointed out forty years ago, the problem is public opinion and values. From a recent survey we learn that of thirteen possible reasons why treasury dollars should be spent for science and technology, "improving health" and "developing energy resources" stand at the top while "discovering new knowledge about man and nature" sits close to the bottom.[32] So cloners must always try to sell themselves as crusaders in the cancer battle. For another thing, the peer review system has certain basic defects. Again, some critics miss the mark, citing the greater need for "egalitarianism" and "improved quality of life" considerations in deciding who gets what perquisites. But from our perspective, the problem is that peer evaluation exalts the cutting edge and the mainstream to the detriment of the quantum leap and the new look.[33] I do not wish to resurrect the great debate in the Supreme Court of the 1940s as to

31. Bernard D. Davis, "Profit Sharing between Professors and the University?" *New England Journal of Medicine* 304 (May 14, 1981): 1234.

32. Walsh, "Public Attitude toward Science," 270.

33. Letter from Daryl E. Chubin published in *Science* 215 (January 15, 1982): 240.

whether scientific innovation features incremental or comprehensive progressions. I do suggest, though, that collegial review, as currently practiced in NIH and NSF backrooms, stacks the deck in favor of the former, if only because, as Max Weber might have said, bureaucratization goes hand-in-hand with rationality and incrementalism is the godchild of both.[34] So peer assessment is a strange beast: it contributes significantly to the posture of funded basic science as constitutional freedom, but it also contributes significantly to funded basic science as a stereotypical, rather than a robust, living constitutional search for truth.

The politics of the reward system has worked even deeper political changes in the nature of scientific inquiry. We are told that new rules of the game spelling out researchers' rights and government's powers are necessary, because the old norms were formulated "when science was a very different social enterprise."[35] How different? Columbia biologist Erwin Chargaff, a pioneer in unraveling the molecular structure of DNA, has called Watson and Crick men of "extreme ignorance," "two pitchmen in search of a helix."[36] What would Chargaff think of Boyer and Gilbert? More to the point, what do cloners think of Boyer and Gilbert? Put bluntly, they are very happy to see them cast their lot with Genentech and Biogen. Professor Chargaff's sense of "purity" could be indulged, perhaps, if the United States were blessed (cursed?) with an aristocracy of Esterhazys, willing and able to fund on the basis of merit alone our cloning Haydns and Mozarts. Maybe that is what the Congress ought to do: create by tax write-off a beehive of twenty-first century Esterhazys, hence saving the integrity of university research from industrial purchase, government strings, and peer evaluation clubiness. Unfortunately, fantasies will not assuage the needs of the recombinant DNA community. The cost of experimentation today is high; geneticists favor overwhelmingly and with good reason an influx of new money from the private sector. The watchword is pragmatism: so far as scientists are concerned, funds can go to either school or scholar depending on whose entrepreneurial skills are sharper. But the watchword is not barbarism: cloners will not accept resources if the free-expression academic laboratory is burned to roast the research pig. What middle grounds are available? What living constitutional encouragements need we contrive in order to protect the university's time-honored place as the preeminent forum for genetic research?

34. And see also Charles Lindblom, "The Science of 'Muddling Through,'" *Public Administration Review* 19 (1959): 79–88.

35. Nelkin, "Intellectual Property," 708.

36. Quoted in Horace Freelund Judson, "Annals of Science (DNA—Part III)," *The New Yorker,* December 11, 1978, 136.

We begin with the assertion that business needs top faculty more than top faculty needs business. That is because the modern-day firm lusts after brainpower, and future Nobelist cloners prefer the pure research climes of Cal Tech and the University of Chicago to Cetus and Genex. Furthermore, we sometimes tend to forget, amidst the Hoechst hoopla, that industry contributes to academic science only 4 percent of the comparable federal expenditure.[37] These realities should establish yet further a context of bargaining between equals. But the proof of that pudding becomes even more evident when we note some of the new arrangements taking root in the physical sciences. For example, the Semiconductor Research Cooperative, a not-for-profit amalgam of computer chip users and makers, has launched a program aimed at funneling as much as $40 million a year into basic academic research of interest to the microelectronics field by 1986. In return, donors would obtain a prepublication look at research findings and royalty-free licenses. It is not clear who will hold patent rights on inventions, but it is clear—and this is of singular importance—that any American company is free to join the consortium and need only contribute funds in proportion to its strength in the industry.[38]

In their zeal to learn from the wise old heads in chemistry and engineering, however, biologists will not find instant political solutions. The fit between pure and applied science is much closer in those two disciplines, so gifts for the former are easier to attract because their consequences in aid of the nation's defense and economic growth commitments are far more obvious. Stanford has established a $20 million Center for Integrated Systems and can count on the Department of Defense to underwrite 40 percent of the cost; Arizona State has received a heavy influx of both corporate and state funds as it seeks to find a place on the high tech map.[39] One of the reasons leading to formation of the Semiconductor Research Cooperative was Japan's inroads into the computer chip market. To this moment, American biotechnology has faced no overseas threat which might incite collective action, but everyone agrees there is plenty of money to be made if firms can only tap the right idea sources.

I propose herewith the blind trust model as an appropriate middle ground. Building on the philosophy of the Economic Recovery Tax Act

37. Barbara J. Culliton, "The Academic-Industrial Complex," *Science* 216 (May 28, 1982): 962.

38. Colin Norman, "Chip Makers Turn to Academe with Offer of Research Support," *Science* 216 (May 7, 1982): 601.

39. Colin Norman, "Electronics Firms Plug into the Universities," *Science* 217 (August 6, 1982): 511-14.

of 1981, the Congress would accord Eli Lilly, Hoffman-La Roche, Genentech, et al. significant tax relief should these companies allocate a fair share of resources, as measured by some objective indicator of prosperity, to an independent nonprofit foundation. That center, employing the standard of review I think best maximizes "goods" over "harms"— collegial evaluation[40]—would support faculty research proposals in the usual fashion. *Individual firms would have no knowledge of their particular beneficiaries, and scholars would have no knowledge of their particular corporate benefactors.* So a Yale professor would be known as a Fellow in the Center for the Study of Cloning rather than as a Du Pont or Monsanto hireling. Patent rights would be held by universities, in a manner consistent with the 1980 statute on federally funded inventions, and royalty-free licenses would be available to donors. The more money contributed, the greater the opportunity for businesses to accumulate licensing options, thus enhancing the prospect that large firms would make large gifts. Patents, or perhaps the grants themselves, could literally be apportioned among sponsors on a percentage basis, through some form of randomization. This element of chance would be offset by the preferential treatment accorded big givers.[41] In any event, it could be the only game in town for commercial outfits if higher education would lobby Congress as effectively as it has lobbied the media to turn college sports into revenue-generating gold mines. And finally, I see no reason why both liberals and conservatives, both partisans and nonpartisans, cannot support such a blueprint with equal enthusiasm.

The central feature of the blind trust is that it guards against conflicts of interest. Even presidents of the United States are considered above

40. There is, however, much that can be done to make peer review a more legitimate embodiment of due process. Justice Thurgood Marshall has argued that those who apply for government jobs are denied fair treatment when their applications are rejected without reason. Board of Regents v. Roth, 408 U.S. 564 (1972), dissenting opinion. One need not concur in his thesis to believe that peer review acquires a gloss of rationality, accountability, and equity when researchers learn why their grant applications fail to attract government subsidy. Therefore, I think reviewers should be put to the test of stating precisely the reasons which have led them to vote against proposals, including proposals submitted to private foundations receiving tax writeoffs. Such a check would place in better perspective and make more visible the tension between mainstream-insiders and new look-outsiders referred to above.

41. It has been suggested that universities tender exclusive licenses to royalty-payers and only nonexclusive licenses to others, a strategy adopted in 1982 by the University of California. Under that approach, though, campus officials must play intermediary between industry and researcher, deciding whom to fund and whom not to fund out of royalties received—assuming that anyone is to be funded. Under my approach, subsidizers need not pay royalties on what they have contributed; nor would exclusivity be a problem if the blind trust concept is maintained.

criticism when they invest in the stock market, provided they do not know where their capital is going and provided the companies involved do not know the chief executive is a shareholder. As applied to the recombinant DNA scene, researchers keep full control over their intellectual property in the fashion of NSF or NIH grant recipients. They may choose to talk or keep quiet about their on-going labors, but neither university-company nor faculty-company contractual arrangements will affect the free flow of information. In short, the patent privilege as employed here poses no threat to free-expression academic interests, and it remains, therefore, a valuable tool for encouraging both science and the useful arts.

This approach is recommended as a viable strategy which at least takes account of the salient political interests, not as a panacea. Several variations on the theme are possible and could well constitute improvements. It may be that many a faculty research quest has descended to the level of sporting event, that the Nobel Prize is now but a sibling of the Heisman Trophy. Universities can neither remake cultures nor alter human ambitions. But they can retain their intellectual integrities and insist that their faculties play by political rules consistent with those integrities, those constitutional values. Such is the policy envisioned here.

Human Genetic Engineering: How to Conceptualize

In November of 1982, the House Subcommittee on Investigations and Oversight, chaired by Congressman Albert Gore, Jr., held three days of hearings on the public policy implications of genetic engineering with human beings. These deliberations were occasioned by the release of a presidential commission report addressing the "social and ethical issues" raised by this especially sensitive brand of cloning.[42] Chief among the commission's recommendations was the creation of a federal agency that would provide "constant vigilance"[43] over the "intended uses of [gene splicing]," with special emphasis on "its diagnostic and therapeutic uses in human beings."[44] The overwhelming body of opinion expressed before Mr. Gore's panel was highly favorable to that initiative. Underneath the surface of commentary could be heard the murmur: "We weren't pre-

42. President's Commission for the Study of Ethical Problems in Medicine and Biomedical and Behavioral Research, *Splicing Life: A Report on the Social and Ethical Issues of Genetic Engineering with Human Beings* (Washington, D.C.: Government Printing Office, 1982).

43. Harold M. Schmeck, Jr., "Ethics Panel Urges Constant Vigilance over Gene-Splicing," *New York Times,* November 16, 1982, 19–20.

44. "Splicing Life" (prepublication copy of the Commission's "Summary of Conclusions and Recommendations"), 1, 4.

pared for the recombinant DNA hullaballoo; we'll be ready next time.'' The question is: Will the nation's political institutions indeed be ready? Mr. Gore promised a swift response in the form of proposed legislation. But even as he must have been reflecting on the various possibilities for a federal presence in the human cloning field,[45] one could see storm clouds of theoretical misdirection on the horizon. For the angle of vision had already assumed a strongly ethical (read: theological) tone, though it should have been clear that the root issues were essentially political (read: constitutional) in their implications.

How did the dialogue over human genetic engineering per se get underway? As was well understood, the NIH guidelines were virtually silent on the matter, simply placing with RAC the authority to review and approve such experiments. Of course, the insertion of alien genes into human subjects would also require permission from IRBs; therefore, even had the NIH gone voluntary on cloning, a mandatory level of censorship would have obtained for these projects. But the lack of attention accorded the issue in recombinant DNA circles, added to the fact that scientists were equivocal about whether these experiments posed special difficulties (as compared to other forms of research with humans),[46] was sufficient to cause concern.

In 1980, President Carter received a petition from three important religious groups—the National Council of Churches, the Synagogue Council of America, and the United States Catholic Conference—expressing anxiety that no federal agency was addressing the "fundamental ethical questions" raised by these procedures.[47] He assigned jurisdiction to his Commission for the Study of Ethical Problems in Biomedical Research, which, under the chairmanship of former Brandeis University president Morris B. Abram, was at that time deeply involved in preparing studies on such formidable topics as the use of life support systems for patients comatose from brain damage. Already, though, the three petitioning organizations had determined for themselves that "those who would play God will be tempted [by the tool of human genetic engineering] as never before" and that no person or group must be permitted to control any unique life forms developed through the new methodology.[48] When we realize that some religious spokesmen, such as Joseph Fletcher, had strongly endorsed these genetic manipulations as a significant force

45. Barbara J. Culliton, "Gore Proposes Oversight of Genetic Engineering," *Science* 218 (December 10, 1982): 1098.

46. *Fed. Reg.* 46 (December 4, 1981): 59390.

47. Schmeck, "Constant Vigilance," 20; "Splicing Life," 1.

48. Charles Austin, "Ethics of Gene Splicing Troubling Theologians," *New York Times,* July 5, 1981, 1.

in the development of "humanhood," while others, fundamentalists, had denounced the new approaches as blasphemous, it seems rather clear that the lines of disputation were not forming on either scientific or political grounds.

The commission's summary of findings commenced with a vote of thanks to petitioners.[49] Their concerns were "well founded," for, in fact, federal administrative agencies and consulting panels had shown little inclination to tackle the ethical questions surrounding gene splicing. The panel then reported its conclusions regarding the "more important issues of substance" which seemed to trouble concerned parties. These observations ran as follows: Fears that genetic engineering might be employed to redesign mankind were "exaggerated"; the techniques, however, were progressing with great rapidity, and they had already begun to demonstrate significant diagnostic and therapeutic value for our species. While the science itself was "a celebration of human creativity" and while the safety questions involved were not unique, "serious efforts to monitor the new [medical] treatment settings" were certainly appropriate. Some goals, such as disease treatment, seemed well within the bounds of recognized practice, but others—the principal example being gene surgery that would initiate evolutionary change and especially where such change involved "enhancing 'normal' people"—deserved "especially close scrutiny." Genetic engineering, then, was not "inherently inappropriate for human use," but a federal agency either in the form of a reconstituted RAC, a reconstituted ethics commission, or some new body was needed to oversee the emerging issues, disseminate information, make recommendations, and possibly frame regulations. That forum, in any event, should be divorced from NIH, and should also represent a cross-section of interests—researchers, lawyers, ethicists, and the general public.

These reflections and proposed actions certainly seemed nontheological, well-balanced, and as steeped in political relevance as a federal government report treating scientific research figured to be. And yet the findings not only failed to cope with the basic policy questions attendant in human genetic engineering but also provided no theroretical focus for the exercise of public functions considered so vital. I submit that the inquiry operated at a marked disadvantage from the outset: it was conducted by an advisory ethics panel at the behest of prominent, powerful religious pressure groups. While the panel was able to steer clear of sectarian trappings and biases, it was unable really to mount a coherent, complete set of politically grounded propositions spelling out what government could do, should do, and why. The central frame of refer-

49. The following paragraph is based on "Splicing Life," 1–5.

ence, of course, is an understanding of relevant constitutional norms and values. I will attempt to enumerate and analyze these.

Perhaps the most fundamental distinction before us separates experimentation from treatment. Experimentation is generally quasi speech; treatment is behavior designed to alleviate suffering, not behavior designed to probe the laws of nature. Experimentation is usually pure research; treatment is, at most, applied science, viz., medicine. Students of ethics have sometimes implied that rules governing human experimentation ought ordinarily to be more rigorous than rules governing human therapy,[50] a thesis I find astounding when measured against the notion of science as expression. I contend that state power is at its zenith when regulating the recombinant DNA procedures of a Martin Cline, procedures which constitute therapeutic (that is, applied) research. I also contend that under the "totality of facts" test, the patient's medical needs must weigh heavier than competing basic science interests, thus providing the state with ample balance-striking regulatory prerogatives. But I think it entirely consistent with the personal privacy analysis outlined in Chapter 2 to submit that in special cases, where, say, the adult patient is terminally ill and tenders informed consent, scientists can perform cloning experiments that are prima facie pure science upon that subject, and in these cases the state lacks constitutional authority to intercede except on a "times, places, manner" basis. *Splicing Life,* the presidential commission's report, fails to address these considerations.

A second fundamental distinction separates the implantation of human genetic material into nonhumans from the implantation of human genetic material into other humans. We may further distinguish the cloning of human DNA in lower organisms from such cloning in higher organisms, for example livestock and primates. Focusing first on the latter subcategories, the use of prokaryotes and the less complex eukaryotes as subjects seems to pose few ethical problems, and perhaps fewer constitutional problems. Yuet Wai Kan's research at the University of California—San Francisco is a case in point. He inserted a defective human beta globin gene into frog egg cells together with a contrived gene for transfer RNA; the result was normal production of the amino acid lysine. This experiment demonstrates the characteristics of a genetic mutation causing beta thalassemia, and is quite clearly constitutionally sheltered under our paradigm.[51] As for cloning human insulin in bacteria

50. See Gray, *Human Subjects,* 2. For purposes of this discussion, I assume that the human subject's constitutional right to be free from unauthorized intrusions into his own body is as viable in the experimental context as in the treatment context. Moreover, I assume that the state's lawmaking power to protect human subjects from these unauthorized invasions is similarly coextensive.

51. Harold M. Schmeck, Jr., "Researchers Correct Defect in Human Gene," *New York Times,* April 26, 1982, 1.

and marketing it as a drug, this is the stuff of applied research and the FDA can place whatever stamps of approval or disapproval on the merchandise that seem warranted.[52]

But what about the use of farm animals as recipients? Late in 1984, Jeremy Rifkin attempted to stop the USDA from inserting human growth hormone genes into sheep and pigs. Here, First Amendment theory seems to draw a line between breeding unique specimens for the sake of expanding our fund of knowledge and breeding these specimens for commercial success. A law, or court order, banning both in one fell swoop might be constitutional as applied to the USDA but certainly not constitutional as applied to the University of Wisconsin.[53] And what about the use of chimpanzees as recipients? Though no such experiments have ever been tried,[54] the prospect seems to have traumatized the commission, which at first condemned the creation of hybrid subhumans as "unacceptable," but later settled on the phrase "very troubling" as an apt description, arguing that studies designed to "improve" these species would only spawn "a group of virtual slaves."[55] The consensus among witnesses testifying at the Gore subcommittee hearings was even more hostile, calling the procedure contrary to our mores and seemingly supporting an outright prohibition.[56] I find these reactions wanting, precisely because they ignore the salient political dimensions. Of course, the ethical standards of this society cannot countenance slavery in any form, whether the bondage be fostered by cloning or by concentration camp. The Constitution speaks directly to this point, proscribing involuntary servitude and vesting with Congress authority to enact necessary and proper legislation that would implement that policy. To be sure, the Thirteenth Amendment was written with *human* slaves in mind, but so long as the statutory definition of those in bondage is reasonable given the language and history of the provision, then that definition is for Congress to formulate. Clearly, it is rational to call quasi humans working at man's behest involuntary servants.

But let us suppose that the only way to determine key biological differences between man and ape is to employ cloning methodologies —more specifically, to recombine into subhuman primates certain human genetic substances. Would such practice be unethical? More to

52. Schmeck, "Constant Vigilance," 20.

53. See my remarks in Chapter 1, note 43.

54. Indeed, as long as we are indulging in fantasy, what about cloning nonhuman genes in fertilized human egg cells? I shall eschew this topic, as scientists consider the likelihoods virtually impossible. Needless to say, volunteer subjects would be rather hard to come by. Harold M. Schmeck, Jr., "Gene Therapy: Scientists Take First Steps toward Curing the 'Incurable' Disorders," *New York Times,* September 8, 1981, C2.

55. Quoted in Schmeck, "Constant Vigilance," 20.

56. Culliton, "Gore Proposes Oversight," 1098.

the point, would such science be per se proscribable? I think not. In fact —and this is the salient matter—a matter eschewed by the president's biomedical ethics commission, the burden of proof should lie with those who would enact content-oriented laws regulating that research. It does seem clear, however, that someone in our constitutional order must have the power to determine the essential attributes of personhood. If a legislature can decide that life ends when the brain stem dies—and I believe, in full accord with another commission report,[57] that it can— then a legislature can decide when a cloned hybrid crosses the line into human existence.[58] If the burden of proof is on those who wish to restrict basic experimentation employing animals, then that burden shifts dramatically when the subject species come to be considered people. We need not enter further this twilight zone, for my argument is simply that without the Constitution, and everything that it has come to stand for, we cannot profitably enter it at all.

Moving back to the context of recombining one person's DNA into another person's DNA,[59] we now confront the level of experimentation which gave the president's commission its most anxious moments, i.e., gene surgery aimed at altering the contents of the subject's germ cells.[60] The procedures could hardly be more controversial, because they implicate man's ability to direct the laws of human evolution. Nor are these operations all of a piece: the least provocative would purge genetic material of the defects that cause such systemic breakdowns as sickle-cell anemia and beta thalassemia, thus making it impossible for these diseases to be inherited; the most provocative would manipulate genetic substances so that offspring could be taller, prettier, healthier, smarter.

57. President's Commission for the Study of Ethical Problems in Medicine and Biomedical and Behavioral Research, *Defining Death: A Report on the Medical, Legal, and Ethical Issues in the Determination of Death* (Washington, D.C.: Government Printing Office, 1981).

58. In Roe v. Wade, 410 U.S. 113 (1973), the Supreme Court said that neither scientists nor anyone else can show when human life begins following conception. At issue there was the pregnant female's right to abort her fetus. Here, life has clearly begun, a woman's freedom to procreate or not to procreate is not at issue, and the only question is whether *human* life has been artificially created out of animal life.

59. Much less controversial than the questions discussed below is gene surgery, which replaces flawed DNA in human somatic cells. The Congressional Office of Technology Assessment has found that this form of therapy should proceed because it involves only the health of the individual recipient. The finding seems eminently sensible. Barbara J. Culliton, "Congress Reports on Gene Therapy," *Science* 226 (December 21, 1984): 1404.

60. The following commentary draws its facts from Constance Holden, "Ethics Panel Looks at Human Gene Splicing," *Science* 217 (August 6, 1982): 516–17. See also "Whether to Make Perfect Humans" (editorial), *New York Times,* July 22, 1982, 22.

To these possibilities, the commission could only propound its "close public scrutiny" test, refusing to recommend any guidelines or legal standards whatever. This circumspection outraged the *New York Times,* which saw the panel caving in to the self-centered values of cloners. The recombinant DNA debate was for them a "big annoyance," intimated *Times* editorialist Nicholas Wade; all gene splicers want to do is hone their craft and play down any further need for public dialogue. I agree with Wade that the commission equivocated inexcusably, but I do not agree with his simplistic assessment of scientists' attitudes and opinions, based as they are on no valid, reliable study of those dynamics. My data have shown that recombinant DNA practitioners are sensitive to the social issues surrounding cloning; that they are prepared to discuss these matters in a reasoned way, if only because they realize they are not islands; and that they expect—indeed, they favor—rules which govern what they do in predictable, uniform style. They do *not* favor, and certainly would not care to enter into discussions concerning, legal constraints based on science fiction scenarios, regulations based on nondemonstrable hazards, and bureaucratic regimens based on political convenience. *There is no evidence that they advocate carte blanche to perform human genetic engineering experiments, and certainly not experiments tampering with the human hereditary process.* It is conceivable, of course, that the president's panel listened with too much deference to scientists who fit Wade's description. But I suspect that the problem is endemic to the larger strategy of the commission, a strategy I have found deficient on political grounds other than the cooptational.

One possible response would be to ban these attempts altogether, to declare man's sperm and egg inviolable. Not even the *New York Times* would go that far. It would throw the subject open to full public debate and wait for the develpment of "an informed understanding"; only then could the science proceed. I am not against debates or understandings. But experimentation showing the workings of genetic flaws and the manner in which they could be deleted from man's DNA bank is pure research, is quasi speech, and is entitled to its appropriate share of constitutional solicitude. The dialogue which the *Times* advocates cannot make of the public a censorship board, through which these inquiries as a form of content must pass. I would even contend that there is a fundamental research component inherent in "enhancement" policy, that, assuming the informed consent rule has been satisfied, scientists retain freedom to find out through cloning methodologies the way in which human intelligence, for example, is carried from generation to generation. Such experiments might very well include germ-cell alteration. By no means does this constitutionalize, or even tolerate, the practice of eugenics.[61] At the level of producing "perfect" offspring, we have left the domain of free expression and entered the world of germinal choice. From the standpoint of the constitutional law paradigm established in Chapter 2, this is the terrain of applied experimentation, and it is a terrain in which the public through its elected representatives should make the necessary value judgments. If the issue boils down to community standards, then there is much to be said for legislation which reserves to parents the option of elimininating those DNA orderings that cause disease but leaves to the forces of nature the fate of all the rest. And yet having said this, we shall see momentarily that my analysis comes up short in the face of constitutional precedent, a predicament which only bears witness to the superficiality of the current dialogue.

There is one final form of genetic engineering we have yet to discuss, and while it is the most audacious cloning exercise ever devised, it received hardly a mention in either the commission's report or the Gore subcommittee's deliberations. I refer to animal asexual reproduction or what is commonly called, in the novelist's vernacular, "human cloning." This methodology, unlike others we have discussed, does not involve the splicing of genes, the recombining of DNA fragments. Rather, the nucleus of an egg cell is removed, and in its stead is inserted the nucleus of a cell from the specimen one wishes to reproduce, viz., clone. The artifact cell is then replanted in the host female, which, thinking it has been impregnated, will give birth to a carbon copy of the donor. Biologists have succeeded in cloning frogs and birds, but they are a long way from cloning pigs much less humans,[62] the plot line in Ira Levin's *The Boys from Brazil* to the contrary notwithstanding. Nonetheless, since we are going to talk about regulating both the manufacture of hybrid human slaves and the manipulation of "intelligence DNA" to "purify" the species, we might as well talk about the constitutional dynamics of regulating this version of human genetic engineering if that exercise would be useful. And, indeed, it is.

Consider a genius musician or playwright and their spouses. They wish their children also to be geniuses—in fact, to be carbon copies of the genius parent. They enlist the support of a cloner and an obstetrician.

61. Here and below I am not talking about laws *forcing* people to bear so-called improved children but, rather, enactments allowing parents to elect that option. I cannot presently contemplate a state interest compelling enough to rationalize compulsory human cloning for enhancement purposes, though the presence of some devastating disorder in the human germ line could prove a justification for involuntary cloning surgery.

62. For descriptions and analyses of these investigations, see Clement L. Markert, "Parthenogenesis, Homozygosity, and Cloning in Mammals," *Journal of Heredity* 73 (November–December 1982): 390–97; Markert, "Cloning Mammals: Current Reality and Future Prospects," *Theriogenology* 21 (January 1984): 60–67.

The task is accomplished in nine months. But the entire process, the entire standard of values, offends the informed citizenry. The state legislature moves to ban such practices. The question is, Can the people's will survive the Constitution? Why wouldn't it?

In the landmark *Griswold* opinion, holding unconstitutional a law making illegal the use of contraceptive devices, Justice Douglas said:

> Would we allow the police to search the sacred precincts of marital bedrooms for telltale signs of the use of contraceptives? The very idea is repulsive to the notions of privacy surrounding the marriage relationship.
> We deal with a right of privacy older than the Bill of Rights—older than our political parties, older than our school system.[63]

The right of privacy is here conceptualized as a substantive liberty, not so sweeping in its terms as scientific inquiry, which has its tap root in free expression theory, but a freedom steeped in community customs and usages, analogous really to the procedural due process which the Court has defended in such important rulings as the oft-cited "stomach pump" decision.[64] So procreative choice, certainly within the bounds of matrimony, is hedged about by a mantle of tradition, of reverence, of privacy, that is embedded in the warp and woof of the Living Constitution. Then why cannot the state proscribe human genetic engineering designed to produce, by germ-line tampering, the super intelligent? Why cannot the people decide even to immunize man's reproductive substances from medical "improvements," deciding, perhaps incorrectly but still as a matter of reasoned judgment, that Pandora's box should remain closed, that the danger of passing on the genetic mutant for sickle-cell anemia is less to be feared than a medical license to clean up, as if it were pollution, defective DNA? And why, I think any rational person might ask, cannot the people ban Ira Levin's "human cloning"? As the president's commission concluded, these are, after all, extraordinary departures from the established order. They certainly are not part and parcel of our Living Constitution.

The answer is *Roe v. Wade.* Our courts have carved out of the right of privacy a subsidiary constitutional right to terminate pregnancy. It makes no difference that the vast majority of states had laws on the books forbidding most abortions as late as 1970. It also makes no dif-

63. Griswold v. Connecticut, 381 U.S. 479, 485-86 (1965).
64. Rochin v. California, 342 U.S. 165 (1952). There, the Court condemned an escapade which saw the California police barge into a private residence (without legal authorization), apprehend the accused, take him into custody, and instruct doctors to pump out his stomach so they could retrieve evidence of morphine, a "search and seizure" which was successfully accomplished. Justice Frankfurter said the entire process "shocks the conscience." Ibid., 172.

ference that these states, in effect, had construed the term "human-hood" to include the fetus. We are compelled to ask: If the state lacks power to forbid *any* abortion unless reasonably related either to the health of the pregnant woman or to the health of the fetus in its advanced stages, then why does the state possess power to ban *any* human genetic engineering procedure designed to further the mother's (or married couples') procreative choice selection, assuming comparable safety standards for human and fetal subjects are provided? That is, if the Court authorizes constitutional blessing for one procreative medical option, then why not others?

At one point during the Gore hearings, a witness made reference to the abortion question. Please, the congressman interrupted, let's not get into that thicket.[65] Unfortunately, the Court has gotten into it already, and the ripple effects for the cloning controversy are marked.[66] It would seem that a critical constitutional image much in need of recombination is the image we have placed on abortion as a private act having enormous political implications for how we govern cloning as scientific innovation.

We learned long ago that information flow was susceptible to precise mathematical measurement and statistical analysis.[67] We now realize that human DNA is an information handbook; that is, hereditary processes are exercises in information flow, providing our species with constitutional parameters.[68] The United States Constitution is information as well, and the power of its message is, in part, a function of information dissemination. It would be no mean feat to apply successfully the notions of informational quantification to the makeup and flow of constitutional values. But it is even more important to realize that human DNA is susceptible to qualitative analysis, that, like our political rules of the game, its contents, while presumptively embodying givens of personal dignity and rationality, are capable of analysis, manipulation, even modest, carefully measured improvement. The game is essentially the same: the common denominator is mature man's essential nature, that sense of freedom balanced against that sense of proportion which features science at its best, politics at its best.

65. This recollection is based on the television coverage provided these discussions.

66. I should state that were I asked to vote for or against laws severely limiting the right to obtain an abortion, I would cast my ballot in the negative. The issue in text, however, is one of *constitutional* interpretation, not statutory decisionmaking. Nor do I say that the Bill of Rights cannot embrace the abortion context. For instance, should the state insist that doctors save the life of the fetus and, in so doing, sacrifice the life of the mother, that policy would, I think, abridge due process.

67. C. E. Shannon and W. Weaver, *The Mathematical Theory of Information* (Urbana: University of Illinois Press, 1949).

68. Harvey Brooks, *The Government of Science* (Cambridge: MIT Press, 1968), 228.

Appendices

Selected Bibliography

Index

Appendix A
Respondent Selection Process

The materials presented in Chapter 4 are taken largely from personal interviews conducted with a random sample of recombinant DNA scientists working at institutions of higher learning. Drawing that sample proved an exceedingly formidable task, posing the sorts of unexpected and unavoidable questions in research validity that are the bane of social science survey investigators. In this appendix, I describe my initial strategy for selecting interviewees, the obstacles I encountered, and the ways in which I sought to work around those pitfalls.

The optimal modus operandi would have been to obtain a master list of all relevant gene splicers, cull from this list by neutral selection process a set of respondents, and then hedgehop around the country interviewing that set. However, the financial costs of such a strategy were clearly prohibitive. Nor could one rely upon a mailed questionnaire to reach this sample, for long, elaborate forms can be administered effectively, I believe, only through face-to-face interaction. Thus, I also ruled out telephone communication as a suitable substitute. In short, it was more appropriate to limit my geographical outreach than it was to fall back on some "staged" means of obtaining the desired rapport and cooperation.

The major regions of the country where cloning is done are, of course, the major regions of academic excellence. Without doubt, the two most prominent clusters are the Boston-Washington corridor and the San Francisco–San Diego coastal belt. Of these, the former is by far the more accessible, as the universities involved are much closer together. In fact, according to data obtained from ORDA in August 1981, there were, at that time, 162 centers of campus gene-splicing experimentation located in the fifty states plus the nation's capitol, and 35, or 22 percent of this total, could be found, by my count, along the Northeast corridor. I decided to make scientists within this sector my universe of recombinant DNA principal investigators.

These best-laid plans received a severe jolt when I discovered that ORDA maintained no complete, up-to-date register of cloning practitioners. I was advised to

contact IBC chairmen—a list of whom Bethesda kindly provided for my use—
and trust that full cooperation would be forthcoming at the campus grassroots
level. I was not sanguine, however, at the prospect of receiving thirty-five sets of
names as a result of either thirty-five letters of request or thirty-five telephone
calls; and I saw no reason why such a protracted process could not be streamlined
effectively by paring my list in half. This I accomplished through an alphabetizing
process, selecting every odd-numbered institution for solicitation. I hoped that
would make available to me gene splicers from eighteen universities, or more than
50 percent of my original campus grouping. Those schools were as follows:

> Albert Einstein College of Medicine
> Boston University Medical School
> Brown University
> University of Delaware
> George Washington University
> Harvard Medical School
> Haverford College
> Hunter College
> Johns Hopkins University
> University of Maryland (Baltimore County)
> Mount Sinai School of Medicine
> New York Medical College
> University of Pennsylvania
> Queen's College
> Rutgers Medical School
> SUNY—Downstate Medical Center
> Temple University
> Yale University

Throughout the months of September and October 1981, I engaged in a series
of telephone exchanges with representatives of these committees, eventually obtaining fifteen rosters of principal investigators. From the beginning, however, it
was clear that Johns Hopkins might not be willing to participate, and finally,
after several attempts at negotiation, legal counsel for the university rejected my
request for information. In the meantime, I had designated by random selection a
back-up institution, and officials at that campus—the New Jersey Medical School
—provided me with the necessary cooperation. Two months having elapsed since
the inception of this process, I made a final effort to obtain data from Boston
University and the University of Delaware, both of which had seemed favorably
disposed to my overtures but had not as yet been forthcoming with information.
These endeavors proved unavailing, and time was now extraordinarily pressing. I
had before me the names of 116 researchers, and I decided, after alphabetization,

to take every sixth entry, for a total of 19, or 16.4 percent, of my universe. I then made contact with the scientists I had pinpointed and found that an overwhelming number were willing to talk to me. In a few cases, face-to-face meetings could not be arranged; in these instances, I selected substitutes randomly from the particular institutions involved. Eventually, all nineteen slots were filled, though one of my choices classified himself as a co-investigator and strongly requested that he and his partner be taken as one. This I agreed to do, and Chapter 4 sometimes records the views of twenty investigators for precisely that reason. I was also able to schedule interviews with seven of the eight IBC panel chairmen whose campuses were represented in the pool, but I was unable to work out a suitable arrangement for meeting with the chief of cloning oversight at the Harvard Medical School.

I wish to thank very much the following recombinant DNA practitioners who shared their thoughts with me:

Sherrill Adams, University of Pennsylvania
Joseph Bertino, Yale University
Gary Cohen, University of Pennsylvania
Jon K. de Riel, Temple University
Peter Dowling, New Jersey Medical School
Roselyn Eisenberg, University of Pennsylvania
Bernard Fields, Harvard Medical School
Bernard Forget, Yale University
Richard Goldberg, Temple University
Gregory Guild, University of Pennsylvania
Ponzy Lu, University of Pennsylvania
William McAllister, Rutgers Medical School
I. George Miller, Yale University
Peter Palese, Mount Sinai School of Medicine
Diane Pratt, Harvard Medical School
Peter Shank, Brown University
Joan Steitz, Yale University
Ian Sussex, Yale University
Yoshinora Takeda, University of Maryland (Baltimore County)
David Ward, Yale University

I would also like to thank the following institutional biosafety committee heads who talked to me:

Edward Adelberg, Yale University
Richard Gethmann, University of Maryland (Baltimore County)
Shalom Hirschman, Mount Sinai School of Medicine
Daniel J. O'Kane, University of Pennsylvania

198

Gerald Shockman, Temple University
William Strohl,[1] College of Medicine and Dentistry of New Jersey
Sumner Twiss, Brown University

1. Head of IBC operations at both Rutgers Medical School and New Jersey Medical School.

Appendix B
Questionnaire for Recombinant DNA Researchers

1. Let me begin by asking about the recombinent DNA research you are conducting. Could you please describe the specific objectives of this research?

2. What role does the recombinant DNA procedure play in that research?

3. How difficult would it be to address the core purposes of your research without being able to employ the recombinant DNA methodology?

4. Would you classify your project as "pure" research or "applied" research?

5. Who is funding this research, and how much money is involved?

6. What physical and biological containment standards, if any, have regulatory agencies placed on your research?

7. (If there are guidelines) What regulatory agencies are these?

8. (If there are guidelines) If you could write the applicable guidelines, what containment standards would you put upon your own research?

9. (If there are guidelines) Overall, would you say the guidelines, as applied to your own work, have been: (a) reasonable, even though perhaps different from what you would prescribe, or (b) unreasonable? (If unreasonable) Why?

10. What role has this institution's biosafety committee played in monitoring your research?

11. Have any points of contention arisen between that committee and yourself as to the conduct of this research, and if so, how were such issues resolved?

12. Have you ever submitted for review any alterations in your research protocol?

13. (If yes) Could you describe what happened, and how long the process took?

14. (If yes) Overall, would you characterize this aspect of the regulatory process as: (*a*) reasonable, even though perhaps different from what you would prescribe, or (*b*) unreasonable? (If unreasonable) Why?

15. With respect to the manner in which the biosafety committee has interpreted its responsibilities where relevant to your work, were such interpretations: (*a*) reasonable, even though perhaps different from your own interpretations, or (*b*) unreasonable? (If unreasonable) Why?

16. Now I'd like to focus on government supervision as a total package of rules, guidelines, and enforcement procedures. Would you say government oversight of your work has been: (*a*) too active, by which I mean overzealous; (*b*) about right; (*c*) not active enough, by which I mean pro forma in nature? Why?

17. Would you rate the manner in which these agencies supervise your research as: (*a*) efficient; (*b*) inefficient? (If inefficient) Why?

18. Some people think the NIH guidelines on recombinant DNA research should be made voluntary. That is, Bethesda could do no more than formulate advisory opinions as to regulatory norms. Do you agree with this?

19. (If yes) Some people think that campus guidelines for recombinant DNA research should be made voluntary. That is, investigators could decide for themselves how to achieve their objectives free from institutional oversight. Do you agree with this?

20. In retrospect, should recombinant DNA researchers have asked the NIH to put mandatory guidelines on their experiments? Why do you think so?

21. Is it appropriate for academic institutions to enter into contracts with business interests for the purpose of subsidizing recombinant DNA research?

22. (If no) Why is it not appropriate?

   (If yes) What "rules of the game" with respect to academic freedom and patent rights should structure these arrangements?

23. Should government put regulatory guidelines on recombinant DNA researchers working in private industry?

24. (If yes) How should these guidelines compare with the NIH regulations? Should they be: (*a*) more strict; (*b*) the same set of guidelines; (*c*) less strict? (If different) How would these guidelines differ?

   (If no) Why not?

25. If you had a choice of having your research funded from among the following three sources, please rank-order which sources you would select: (*a*) private industry; (*b*) the federal government; (*c*) your home institution, but under conditions where that institution relies on royalties from patents developed by faculty working in your research area.

26. Why do you rate these options as you do?

27. Let's assume that your research bears fruit. Will you consider your contributions to be your own private property, or will you consider them to be a part of the marketplace of ideas, accessible to all? (If "a little bit of both") How would you separate these out?

28. Should academic institutions permit their recombinant DNA researchers to work up arrangements with private corporations for the purpose of developing and marketing the products of their labors?

29. There has been some debate lately about whether scientific research is protected by the First Amendment of the U.S. Constitution. Do you think scientific inquiry is entitled to such protection?

30. (If yes) What about recombinant DNA procedures and the Constitution? Do you think the First Amendment protects the recombinant DNA research you are doing?

31. The Supreme Court has recently said that the individual possesses a constitutional right of privacy. Do you think the laboratory environment is protected by this right of privacy?

32. (If yes) Do you think the recombinant DNA research you are doing is protected by this right of privacy?

33. Let's assume for a moment that you subsidized your own research. Given your view of scientific inquiry and constitutional rights, to what extent could government pass laws regulating the recombinant DNA research you do?

34. Now let's say that the government *is* funding your research. Some people say that if government is paying the scientist's way, then it can attach whatever strings it desires on the scientist's research. Do you think government has the constitutional power to attach any strings it wants on your recombinant DNA research?

35. (If no) What strings can't government attach?

36. (If no) In your judgment, has government ever attached such strings on your research?

37. Do you think government has the constitutional power to fund genetic research, but then refuse to allow any money to be used for recombinant DNA research?

38. The U.S. Constitution makes it impermissible for government to take either liberty or property without due process of law. With respect to the *procedures* which government agencies including your institutional biosafety committee have placed upon you, can you report any instances when due process has been ignored? How?

39. Putting your own work to one side, do you believe government agencies have at times violated the constitutional rights of scientists to perform recombinant DNA research? What instances have you in mind?

40. With respect to political party identification, would you call yourself: (*a*) a Republican; (*b*) a Democrat; (*c*) an independent; (*d*) other? (If "other") What "other" is this?

41. Is there any connection between your choice of party affiliation and the fact that you do the work you do? What common attitude or linkage is this?

42. With respect to political ideology, would you call yourself: (*a*) a liberal; (*b*) a conservative; (*c*) a middle-of-the-roader; (*d*) other? (If "other") What "other" is this?

43. Is there any connection between your political ideology and the fact that you do the work you do? What common attitude or linkage is this?

Appendix C
Questionnaire for Institutional Biosafety Committee Chairmen

1. Let me begin by focusing on the makeup of the ——— biosafety committee. How many committee members are there?

2. Could you define the composition of the committee from the standpoint of professional training? How many members are: (*a*) recombinant DNA researchers; (*b*) other natural or physical scientists; (*c*) social scientists; (*d*) medical doctors; (*e*) lawyers; (*f*) full-time administrators; (*g*) community representatives?

3. As committee chairman, do you possess any special expertise or training with respect to recombinant DNA research? What is that?

4. Were committee members as a group provided with any special training regarding recombinant DNA research?

5. How would you describe your activities on the board as compared with those of your peers? That is, in your role as chairman, do you perform any special tasks in the decisionmaking process?

6. Approximately how often has the committee refused to approve a proposed recombinant DNA project? Could you give me a ballpark percentage?

7. (If there have been instances) In approximately what percentage of these cases did the principal investigator satisfy the committee by making appropriate modifications?

8. (If there have been instances) With regard to these cases, what was the principal reason why the committee felt modifications were in order?

9. (If some proposals *never* went through) Why did the other proposals never receive approval?

10. Given an instance of committee rejection, does the researcher have a right of appeal? (If yes) To whom is such appeal taken?

11. (Where there have been rejections, and where there is a right of appeal) Approximately how often have appeals been lodged, and how were these resolved?

12. Does your committee take formal votes on recombinant DNA proposals? (If yes) Is there a record kept of these votes? Does majority vote suffice?

13. How would you describe the level of agreement at committee meetings? Would you say there is generally: (a) no disagreement; (b) some disagreement; (c) considerable disagreement?

14. In looking at the difference of opinion in your committee, would you say the "losing" side in a controversy tends to represent a force for greater regulation or for lesser regulation?

15. Are principal investigators permitted to attend meetings at which their proposals come up for discussion?

16. Over and above the authorship of regulatory guidelines, have NIH personnel in Washington played any role in supervising or overseeing such research at this institution? (If yes) What role has this been?

17. How often have recombinant DNA researchers at this institution submitted for review any alterations in their research protocols? Would you say such requests occur frequently or infrequently?

18. (If there were some) Could you describe what happened in these cases, and how long the process took?

19. Does the committee have any procedures to monitor the progress of recombinant DNA research? (If yes) What are these procedures?

20. (Where there is monitoring) Have any points of contention arisen between the committee and a recombinant DNA researcher as a result of such monitoring? (If yes) How were these resolved?

21. One of the projects over which you have jurisdiction is ———'s study. Would you classify his (her) investigations as "pure" research or "applied" research?

22. As you view the regulatory guidelines your committee enforces, do you think they have been: (a) too strict; (b) about right; (c) not strict enough?

23. In retrospect, have the committee's interpretations of these guidelines not been perhaps what they should have been? More particularly, can you point to any errors in judgment that may have been committed?

24. In retrospect, has the supervision of recombinant DNA research on this campus been as efficient as it might have been? More particularly, can you point to aspects of the regulatory process which have shown themselves to be inefficient?

25. Some people think the NIH guidelines on recombinant DNA research should be made voluntary. That is, Bethesda could do no more than formulate advisory opinions as to regulatory norms. Do you agree with this?

26. (If yes) Some people think that campus guidelines for recombinant DNA research should be made voluntary. That is, investigators could decide for themselves how to achieve their objectives free from institutional oversight. Do you agree with this?

27. In characterizing the relationship between your committee and your home institution, would you say that: (a) the committee enjoys a very wide range of discretion; (b) there is pretty careful supervision of what your committee is doing; (c) the committee must cope with pressures from the larger institutional community?

28. In retrospect, should recombinant DNA researchers have asked NIH to put mandatory guidelines on their experiments? Why do you think so?

29. Is it appropriate for academic institutions to enter into contracts with business interests for the purpose of subsidizing recombinant DNA research?

30. (If no) Why is it not appropriate?

    (If yes) What "rules of the game" with respect to academic freedom and patent rights should structure these arrangements?

31. Should government put regulatory guidelines on recombinant DNA researchers working in private industry?

32. (If yes) How should these guidelines compare with the NIH regulations? Should they be: (a) more strict; (b) the same set of guidelines; (c) less strict? (If different) How would these guidelines differ?

    (If no) Why not?

33. Should academic institutions permit their recombinant DNA researchers to work up arrangements with private corporations for the purpose of developing and marketing the products of their labors?

34. There has been some debate lately about whether scientific research is protected by the First Amendment of the U.S. Constitution. Do you think scientific inquiry is entitled to such protection?

35. (If yes) What about recombinant DNA procedures and the Constitution? Do you think the First Amendment protects the recombinant DNA research your committee regulates?

36. As you may know, the Supreme Court has recently said that the individual possesses a constitutional right of privacy. Do you think the laboratory environment is protected by this right of privacy?

37. (If yes) Do you think the recombinant DNA research your committee regulates is protected by this right of privacy?

38. Let's assume for a moment that you were doing recombinant DNA research and that you were subsidizing your own investigations. Given your view of scientific inquiry and constitutional rights, to what extent could government pass laws regulating your research?

39. Now let's say that the government *was* funding your recombinant DNA research. Some people say that if government is paying the scientist's way, then it can attach whatever strings it desires on the scientist's research. Do you think government has the constitutional power to attach any strings it wants to such research?

40. (If no) What strings can't government attach?

41. (If no) In your judgment, has government ever attached such strings to research which your committee regulates?

42. Do you think government has the constitutional power to fund genetic research, but then refuse to allow any money to be used for recombinant DNA research?

43. The U.S. Constitution makes it impermissible for government to take either liberty or property without due process of law. With respect to the *procedures* which government agencies including your committee have placed on recombinant DNA research at this institution, do you think there have been instances when due process has not been provided? How?

44. Putting research at this institution to one side, do you believe government agencies have at times violated the constitutional rights of scientists to perform recombinant DNA research? What instances have you in mind?

45. With respect to political party identification, would you call yourself: (*a*) a Republican; (*b*) a Democrat; (*c*) an independent; (*d*) other? (If "other") What "other" is this?

46. Is there any connection between your choice of party affiliation and the fact that you do the work you do? What common attitude or linkage is this?

47. With respect to political ideology, would you call yourself: (*a*) a liberal; (*b*) a conservative; (*c*) a middle-of-the-roader; (*d*) other? (If "other") What "other" is this?

48. Is there any connection between your political ideology and the fact that you do the work you do? What common attitude or linkage is this?

Selected Bibliography

Books and Essays Therein

Barber, Bernard, et al. *Research on Human Subjects.* New York: Sage, 1973.

Brooks, Harvey. *The Government of Science.* Cambridge: MIT Press, 1968.

Bush, Vannevar. *Science—The Endless Frontier.* Washington, D.C.: Government Printing Office, 1945.

Carmen, Ira H. *Power and Balance.* New York: Harcourt Brace Jovanovich, 1978.

Chafee, Zechariah, Jr. *Freedom of Speech.* New York: Harcourt, Brace and Howe, 1920.

Cohen, I. Bernard. *Franklin and Newton.* Philadelphia: *American Philosophical Society,* 1956.

Crowther, J. G. *Famous American Men of Science.* Freeport, N.Y.: Books for Libraries, 1969.

Dupré, J. Stefan and Sanford A. Lakoff. *Science and the Nation.* Englewood Cliffs: Prentice-Hall, 1962.

Dupree, A. Hunter. *Science in the Federal Government.* Cambridge: Harvard University Press, 1957.

Emerson, Thomas I. *The System of Freedom of Expression.* New York: Vintage, 1970.

Fred, E. B. *The Role of the Wisconsin Alumni Research Foundation in the Support of Research at the University of Wisconsin.* Madison: WARF, 1973.

Fudenberg, H. H., and V. L. Melnick, eds. *Biomedical Scientists and Public Policy.* New York: Plenum, 1978.

Gray, Bradford H. *Human Subjects in Medical Experimentation.* New York: Wiley, 1975.

Grobstein, Clifford. *A Double Image of the Double Helix.* San Francisco: Freeman, 1979.

Haberer, Joseph. *Politics and the Community of Science.* New York: Van Nostrand Reinhold, 1969.

Inlow, E. Burke. *The Patent Grant.* Baltimore: Johns Hopkins Press, 1950.

Krimsky, Sheldon. *Genetic Alchemy: The Social History of the Recombinant DNA Controversy.* Cambridge: MIT Press, 1982.

Leontief, Wassily W. "Introduction." In Leonard S. Silk, *The Research Revolution*. New York: McGraw-Hill, 1960.

Lowi, Theodore J. *The End of Liberalism*. Chicago: Norton, 1969.

Meiklejohn, Alexander. *Free Speech and Its Relation to Self-Government*. New York: Harper, 1948.

Miller, Jon D., Kenneth Prewitt, and Robert Pearson. *The Attitudes of the U.S. Public toward Science and Technology*. Chicago: National Opinion Research Center, University of Chicago, 1980.

Morgan, J., and W. J. Whelan, eds. *Recombinant DNA and Genetic Experimentation*. New York: Pergamon, 1979.

National Academy of Sciences. *Research with Recombinant DNA: An Academy Forum*. Washington, D.C.: NAS, 1977.

Nelkin, Dorothy. *The University and Military Research*. Ithaca: Cornell University Press, 1972.

Padover, Saul K., ed. *A Jefferson Profile*. New York: Day, 1956.

Price, Don K. *The Scientific Estate*. Cambridge: Harvard University Press, 1965.

Shapiro, Martin. *Freedom of Speech*. Englewood Cliffs: Prentice-Hall, 1966.

Szybalski, W. "Much Ado about Recombinant DNA Regulations." In H. H. Fudenberg and V. L. Melnick, eds., *Biomedical Scientists and Public Policy*. New York: Plenum, 1978.

Szybalski, W. "Chairman's Introduction." In J. Morgan and W. J. Whelan, eds., *Recombinant DNA and Genetic Experimentation*. New York: Pergamon, 1979.

Watson, James D. *The Double Helix*. New York: Atheneum, 1968.

Watson, James D., and John Tooze. *The DNA Story: A Documentary History of Gene Cloning*. San Francisco: Freeman, 1981.

Wood, Robert C. "Scientists and Politics: The Rise of an Apolitical Elite." In Robert Gilpin and Christopher Wright, eds., *Scientists and Rational Policy Making*. New York: Columbia University Press, 1964.

Court Cases

Arnett v. Kennedy, 416 U.S. 134 (1974).

Barenblatt v. United States, 360 U.S. 109 (1959).

Blount v. Rizzi, 400 U.S. 410 (1971).

Bob Jones University v. United States, no. 81–3 (1983).

Buckley v. Valeo, 424 U.S. 1 (1976).

Cantwell v. Connecticut, 310 U.S. 296 (1940).

Carey v. Brown, 447 U.S. 445 (1980).

Carey v. Population Services, 431 U.S. 678 (1977).

Chaplinsky v. New Hampshire, 315 U.S. 568 (1942).

Chicago Police Department v. Mosely, 408 U.S. 92 (1972).

Civil Service Commission v. National Association of Letter Carriers, 413 U.S. 548 (1973).

Cousins v. Wigoda, 319 U.S. 477 (1975).

Cox v. Louisiana, 379 U.S. 559 (1965).

Cox v. New Hampshire, 312 U.S. 569 (1941).

Cuno Corp. v. Automatic Devices Corp., 314 U.S. 84 (1941).

Dennis v. United States, 341 U.S. 494 (1951).

Erznoznik v. Jacksonville, 422 U.S. 205 (1975).

Freedman v. Maryland, 380 U.S. 51 (1965).

Goldberg v. Kelly, 397 U.S. 254 (1970).

Goss v. Lopez, 419 U.S. 565 (1975).

Griswold v. Connecticut, 381 U.S. 479 (1965).

Harris v. McRae, 448 U.S. 297 (1980).

Healy v. James, 408 U.S. 169 (1972).

Keyishian v. Board of Regents, 385 U.S. 589 (1967).

Kingsley Pictures v. Regents, 360 U.S. 684 (1959).

Lamont v. Postmaster General, 381 U.S. 301 (1956).

McAuliffe v. Mayor of New Bedford, 155 Massachusetts 216 (1892).

McKeiver v. Pennsylvania, 403 U.S. 528 (1971).

Marconi Wireless Co. v. U.S., 320 U.S. 1 (1942).

Martin v. Struthers, 319 U.S. 141 (1943).

Mathews v. Eldridge, 424 U.S. 319 (1976).

Meyer v. Nebraska, 262 U.S. 390 (1923).

Miller v. California, 413 U.S. 15 (1973).

Milwaukee Publishing Co. v. Burleson, 255 U.S. 407 (1921).

Palko v. Connecticut, 302 U.S. 319 (1937).

Planned Parenthood v. Danforth, 428 U.S. 52 (1976).

Powell v. Alabama, 287 U.S. 45 (1932).

Rochin v. California, 342 U.S. 165 (1952).

Roe v. Wade, 410 U.S. 113 (1973).

Roth v. United States, 354 U.S. 476 (1957).

Schenck v. United States, 249 U.S. 47 (1919).

Shapiro v. Thompson, 394 U.S. 618 (1969).

Shelton v. Tucker, 364 U.S. 479 (1960).

Sherbert v. Verner, 374 U.S. 398 (1963).

Shuttlesworth v. Birmingham, 394 U.S. 147 (1969).

Stanley v. Georgia, 394 U.S. 557 (1969).

Sweezy v. New Hampshire, 354 U.S. 234 (1957).

Thornhill v. Alabama, 310 U.S. 88 (1940).

Tinker v. Des Moines School District, 393 U.S. 503 (1969).

United States v. O'Brien, 391 U.S. 367 (1968).

Virginia State Board v. Virginia Citizens Consumer Council, 425 U.S. 748 (1976).

Wyman v. James, 400 U.S. 309 (1971).

Periodicals and Journal Articles

Abelson, John. "Recombinant DNA: Examples of Present-Day Research." *Science* 196 (April 8, 1977): 159–60.

Barnum, David G. "Decision Making in a Constitutional Democracy: Policy Formation in the Skokie Free Speech Controversy." *Journal of Politics* 44 (May 1982): 480–508.

Becker, Frank. "Law vs. Science: Legal Control of Genetic Research." *Kentucky Law Journal* 65 (1977): 880–94.

Bereano, Philip L. "Institutional Biosafety Committees and the Inadequacies of Risk Regulation." *Science, Technology, and Human Values* 9 (Fall 1984): 16–34.

Bork, Robert. "Neutral Principles and Some First Amendment Problems." *Indiana Law Journal* 47 (1971): 1–35.

Carmen, Ira H. "The Constitution in the Laboratory: Recombinant DNA Research as 'Free Expression.'" *Journal of Politics* 43 (1981): 737–62.

Cohen, Stanley N. "Recombinant DNA: Fact and Fiction." *Science* 195 (February 18, 1977): 654–57.

Crittenden, Ann. "Industry's Role in Academia." *New York Times,* July 22, 1981, D1, D11.

Crittenden, Ann. "Universities' Accord Called Research Aid." *New York Times,* September 12, 1981, 32.

Culliton, Barbara J. "Biomedical Research Enters the Marketplace." *New England Journal of Medicine* 304 (May 14, 1981): 1195–1201.

Culliton, Barbara J. "Pajaro Dunes: The Search for Consensus." *Science* 216 (April 9, 1982): 155–58.

Culliton, Barbara J. "The Academic-Industrial Complex." *Science* 216 (May 28, 1982): 960–62.

Culliton, Barbara J. "The Hoechst Department at Mass General." *Science* 216 (June 11, 1982): 1200–1203.

Davis, Bernard D. "Profit Sharing between Professors and the University?" *New England Journal of Medicine* 304 (May 14, 1981): 1232–35.

Delgado, Richard, and David R. Millen. "God, Galileo, and Government: Toward Constitutional Protection for Scientific Inquiry." *Washington Law Review* 53 (1978): 349–404.

Delgado, Richard, et al. "Can Science Be Inopportune? Constitutional Validity of Governmental Restrictions on Race-IQ Research." *UCLA Law Review* 31 (1983): 128–225.

Emerson, Thomas I. "Toward a General Theory of the First Amendment." *Yale Law Journal* 72 (1963): 877–956.

Feder, Barnaby J. "Gene-Splicing Licensed." *New York Times,* August 19, 1981, D1–D2.

Fox, Jeffrey L. "Rifkin Takes Aim at USDA Research." *Science* 226 (October 19, 1984): 321.

Goggin, Malcolm L. "The Life Sciences and the Public: Is Science Too Important to Be Left to the Scientists?" *Politics and the Life Sciences* 3 (August 1984): 28–40.

Golden, Frederic. "Shaping Life in the Lab." *Time,* March 9, 1981, 50–59.

Graham, Loren R. "Concerns about Science and Attempts to Regulate Inquiry." *Daedalus* 107 (Spring 1978): 1–21.

Green, Harold P. "The Boundaries of Scientific Freedom." *Harvard Newsletter on Science, Technology, and Human Values,* June 1977, 17–21.

Grobstein, Clifford. "The Recombinant-DNA Debate." *Scientific American* 237 (July 1977): 22–33.

Holden, Constance. "Ethics Panel Looks at Human Gene Splicing." *Science* 217 (August 6, 1982): 516–17.

Jonas, Hans. "Freedom of Scientific Inquiry and the Public Interest." *Hastings Center Report* 6 (August 1976): 15–17.

Judson, Horace Freelund. "Annals of Science (DNA: 1–III)." *The New Yorker,* November 27, 1978, 46–124; December 4, 1978, 89–191; December 11, 1978, 122–96.

Knox, Richard A. "Secrecy and Questions in Gold-Rush Atmosphere." *Boston Globe,* September 15, 1981, 1, 8–9.

Krimsky, Sheldon. "Local Monitoring of Biotechnology: The Second Wave of Recombinant DNA Laws." *Recombinant DNA Technical Bulletin* 5 (June 1982): 79–84.

Lappé, Marc, and Patricia Archbold Martin. "The Place of the Public in the Conduct of Science." *Southern California Law Review* 51 (1978): 1535–54.

Lewin, Roger. "Judge's Ruling Hits Hard at Creationism." *Science* 215 (January 22, 1982): 381–84.

Llewellyn, Karl N. "The Constitution as an Institution." *Columbia Law Review* 34 (1934): 1–40.

Markert, Clement L. "Parthenogenesis, Homozygosity, and Cloning in Mammals." *Journal of Heredity* 73 (November–December 1982): 390–97.

Markert, Clement L. "Cloning Mammals: Current Reality and Future Prospects." *Theriogenology* 21 (January 1984): 60–67.

Milewski, Elizabeth. "Report of the IBC Chairpersons' Meeting." *Recombinant DNA Technical Bulletin* 4 (April 1981): 19–27.

Nelkin, Dorothy. "Intellectual Property: The Control of Scientific Information." *Science* 216 (May 14, 1982): 704–8.

Norman, Colin. "MIT Agonizes Over Links with Research Unit." *Science* 214 (October 23, 1981): 416–17.

Norman, Colin. "Electronics Firms Plug into the Universities." *Science* 217 (August 6, 1982): 511–14.

Pillar, Charles. "DNA—Key to Biological Warfare?" *The Nation,* December 10, 1983, 596–601.

Robertson, John A. "The Scientist's Right to Research: A Constitutional Analysis." *Southern California Law Review* 51 (1978): 1203–79.

Rosenbaum, Robert A., et al. "Academic Freedom and the Classified Information System." *Science* 219 (January 21, 1983): 257–59.

Schmeck, Harold M., Jr. "Gene Therapy: Scientists Take First Steps Toward Curing the 'Incurable' Disorders." *New York Times,* September 8, 1981, C1–C2.

Schmeck, Harold M., Jr. "Ethics Panel Urges Constant Vigilance over Gene-Splicing." *New York Times,* November 16, 1982, 19–20.

Stetten, D. "Freedom of Enquiry." *Genetics* 81 (1975): 415–25.

Stetten, D. "Valedictory by the Chairman of the NIH Recombinant DNA Molecule Program Advisory Committee." *Gene* 3 (1978): 265–68.

"Summary Statement of the Asilomar Conference on Recombinant DNA Molecules, May 1975." *Science* 188 (June 6, 1975): 991–94.

Sun, Marjorie. "NIH Ponders Pitfalls of Industrial Support." *Science* 213 (July 3, 1981): 113–14.

Sun, Marjorie. "Cline Loses Two NIH Grants." *Science* 214 (December 11, 1981): 1220.

Sun, Marjorie. "NIH Developing Policy on Misconduct." *Science* 216 (May 14, 1982): 711–12.

Swisher, Carl B. "The Supreme Court and 'The Moment of Truth.'" *American Political Science Review* 54 (December 1960): 879–86.

Vermeulen, Michael. "Harvard Passes the Buck: The DNA Affair." *TWA Ambassador,* January 1982, 41–57.

Wade, Nicholas. "Gene-Splicing: Cambridge Citizens OK Research But Want More Safety." *Science* 195 (January 21, 1977): 268–69.

Wade, Nicholas. "DNA: Law, Patents, and a Proselyte." *Science* 195 (February 25, 1977): 762.

Wade, Nicholas. "Recombinant DNA: New York State Ponders Action to Control Research." *Science* 194 (November 12, 1977): 705–6.

Wade, Nicholas. "Gene Therapy Caught in More Entanglements." *Science* 212 (April 3, 1981): 24–25.

Wade, Nicholas. "Gold Pipettes Make for Tight Lips." *Science* 212 (June 19, 1981): 1368.

Walsh, John. "DOD Springs Surprise on Secrecy Rules." *Science* 224 (June 8, 1984): 1081.

Watson, James D. "In Defense of DNA." *New Republic,* June 25, 1977, 11–14.

Wechsler, Herbert. "Toward Neutral Principles of Constitutional Law." *Harvard Law Review* 73 (November 1959): 1–35.

Zieg, Janine, et al. "Recombinational Switch for Gene Expression." *Science* 196 (April 8, 1977): 170–72.

Public Documents

"Guidelines for the Use of Recombinant DNA Molecule Technology in the City of Cambridge, January 1977." Reprinted as Appendix 4 in Clifford Grobstein, *A Double Image of the Double Helix.* San Francisco: Freeman, 1979.

National Institutes of Health. Recombinant DNA Advisory Committee. Minutes of Meetings (April 1981–February 1982).

National Science Board. *Basic Research in the Mission Agencies: Agency Perspectives on the Conduct and Support of Basic Research.* NSB-7-1. Washington, D.C.: Government Printing Office, 1978.

National Science Foundation. *Federal Organization for Scientific Activities, 1962.* NSF 62-37. Washington, D.C.: Government Printing Office, 1963.

National Science Foundation. *Federal Funds for Research, Development, and Other Scientific Activities.* NSF 77-301, XXV. Washington, D.C.: Government Printing Office, 1976.

Office of Management and Budget. "Circular A-21—Cost Principles for Educational Institutions." *Federal Register* 44 (1979): 12368-12380.

President's Commission for the Study of Ethical Problems in Medicine and Biomedical and Behavioral Research. *Splicing Life: A Report on the Social and Ethical Issues of Genetic Engineering with Human Beings.* Washington, D.C.: Government Printing Office, 1982.

"Regulating the Use of Recombinant DNA Technology." Boston Ordinances of 1981, chap. 12, doc. 50-1981.

U.S. Congress. House of Representatives. Committee on Science and Technology. *Science Policy Implications of DNA Recombinant Molecule Research.* Hearings before the Subcommittee on Science, Research, and Technology. 95th Cong., Report No. 24. Washington, D.C.: Government Printing Office, 1977.

U.S. Congress. Senate. Committee on Commerce, Science, and Transportation. *Regulation of Recombinant DNA Research.* Hearings before the Subcommittee on Science, Technology, and Space. 95th Cong. Washington, D.C.: Government Printing Office, 1978.

U.S. Department of Health, Education, and Welfare. National Institutes of Health. *Recombinant DNA Research: Documents Relating to "NIH Guidelines for Research Involving Recombinant DNA Molecules,"* Vols. 1-6 (1976-1981).

U.S. Department of Health, Education, and Welfare. National Institutes of Health. "Guidelines for Research Involving Recombinant DNA Molecules, June 1976." *Federal Register* 41, No. 131 (July 1976): 27911-27922.

U.S. Department of Health, Education, and Welfare. National Institutes of Health. "Proposed Revised Guidelines." *Federal Register* 43, No. 146 (July 28, 1978): 33042-33178.

U.S. Department of Health, Education, and Welfare. National Institutes of Health. "Revised Guidelines." *Federal Register* 43, No. 247 (December 22, 1978): 60080-60130.

U.S. Department of Health and Human Services. *NIH Extramural Programs.* NIH Pub. No. 80-33, July 1980.

U.S. Department of Health and Human Services. National Institutes of Health. "Recombinant DNA Research; Actions Under Guidelines." *Federal Register* 47, No. 77 (April 21, 1982): 17172-17176.

Index

217

COMPOSED BY THE COMPOSING ROOM, INC.
APPLETON, WISCONSIN
MANUFACTURED BY CUSHING MALLOY, INC.
ANN ARBOR, MICHIGAN
TEXT IS SET IN TIMES ROMAN, DISPLAY LINES IN KABEL

Library of Congress Cataloging-in-Publication Data
Carmen, Ira H.
Cloning and the Constitution.
Bibliography: pp. 209–215.
Includes index.
1. Genetic engineering—Law and legislation—
United States. 2. Genetic engineering—Government
policy—United States. 3. Molecular cloning—
Government policy—United States. I. Title.
KF3827.G4C37   1985   344.73′095   85-40363
ISBN 0-299-10340-4   347.30495